BUS

Cover, Title Page, and Dividers

Illiterate Medieval worshippers could only be instructed by the churches' stained glass windows. This book is surely meant for persons who can read but who might have forgotten how to use their eye. They will be urged to treasure and train them, without respite. Each picture from the cover, title page, and section dividers is reproduced in miniature in this primer, and credited; the Guide to Topics refers to chapters that discuss it.

Cauliflower dish. Reflection in a horizontal mirror yields a surface drawn on the same principles as a curve Takagi drew in 1900. The most primitive cartoon of a mountain, this collection of bumps upon bumps upon bumps is *not* self-similar (like the cauliflower) but self-affine: the details become increasingly peaked as one zooms in. *Credit*: B.B. Mandelbrot & R. Gagne.

Mountain. The Earth's relief predates everything else in this book, even cauliflowers, but this mountain is only a forgery: a self-affine fractal gestated in a computer and born on a monitor. *Credit*: F.Kenton Musgrave & Pandromeda.com., 2000.

Mandelbrot set. An icon of fractal geometry rising modestly behind the fractal mountains, source of many open questions, a rare example in mathematical physics of instant gratification: a simple formula producing an extraordinarily complicated fractal that enchants and baffles scientists and everyone else. *Credit*: N. Eisenkraft, 2001.

Cow's head. The art of the title page is not a caricature of the head of an animal. It is a self-affine fractal produced by an iterated function system, as part of a student's project in a fractal geometry course at Lehigh University. Methods like this use the fractal character of natural scenes and are of significant use in image compression. *Credit*: Sarah Goode & Don Davis, 1995.

African village. This schematic of the real ruined village of Ba-Ila in Tanzania demonstrates that an intuitive awareness of fractality predates writing and organized mathematics. It seems that, at one time or another, every culture has used fractals in decoration and architecture. *Credit*: R. Eglash, 1999.

DLA coral. This is a product of neither nature nor art but of statistical physics. It is a fractal construction that imitates the process of diffusion-limited aggregation, hence its initials. It is a prototype of the fractal nature of spontaneous growth, a rare example in physics of instant gratification, insofar as the path is very short from a simple formula to impossibly difficult questions. *Credit*: B.B. Mandelbrot & H. Kaufmann, 1994.

Fractals, Graphics,
and Mathematics Education

© 2002 by
The Mathematical Association of America (Incorporated)
Library of Congress Catalog Card Number 2001097388

ISBN 0-88385-169-5

Printed in the United States of America

Current printing (last digit):
10 9 8 7 6 5 4 3 2 1

Fractals, Graphics, and Mathematics Education

Michael Frame and Benoit B. Mandelbrot

editors

Foreword by Lynn Arthur Steen

Published and Distributed by
The Mathematical Association of America

The MAA Notes Series, started in 1982, addresses a broad range of topics and themes of interest to all who are involved with undergraduate mathematics. The volumes in this series are readable, informative, and useful, and help the mathematical community keep up with developments of importance to mathematics.

MAA Notes

11. Keys to Improved Instruction by Teaching Assistants and Part-Time Instructors, *Committee on Teaching Assistants and Part-Time Instructors, Bettye Anne Case,* Editor.

13. Reshaping College Mathematics, *Committee on the Undergraduate Program in Mathematics, Lynn A. Steen,* Editor.

14. Mathematical Writing, by *Donald E. Knuth, Tracy Larrabee, and Paul M. Roberts.*

16. Using Writing to Teach Mathematics, *Andrew Sterrett,* Editor.

17. Priming the Calculus Pump: Innovations and Resources, *Committee on Calculus Reform and the First Two Years,* a subcommittee of the Committee on the Undergraduate Program in Mathematics, *Thomas W. Tucker,* Editor.

18. Models for Undergraduate Research in Mathematics, *Lester Senechal,* Editor.

19. Visualization in Teaching and Learning Mathematics, *Committee on Computers in Mathematics Education, Steve Cunningham and Walter S. Zimmermann,* Editors.

20. The Laboratory Approach to Teaching Calculus, *L. Carl Leinbach et al.,* Editors.

21. Perspectives on Contemporary Statistics, *David C. Hoaglin and David S. Moore,* Editors.

22. Heeding the Call for Change: Suggestions for Curricular Action, *Lynn A. Steen,* Editor.

24. Symbolic Computation in Undergraduate Mathematics Education, *Zaven A. Karian,* Editor.

25. The Concept of Function: Aspects of Epistemology and Pedagogy, *Guershon Harel and Ed Dubinsky,* Editors.

26. Statistics for the Twenty-First Century, *Florence and Sheldon Gordon,* Editors.

27. Resources for Calculus Collection, Volume 1: Learning by Discovery: A Lab Manual for Calculus, *Anita E. Solow,* Editor.

28. Resources for Calculus Collection, Volume 2: Calculus Problems for a New Century, *Robert Fraga,* Editor.

29. Resources for Calculus Collection, Volume 3: Applications of Calculus, *Philip Straffin,* Editor.

30. Resources for Calculus Collection, Volume 4: Problems for Student Investigation, *Michael B. Jackson and John R. Ramsay,* Editors.

31. Resources for Calculus Collection, Volume 5: Readings for Calculus, *Underwood Dudley,* Editor.

32. Essays in Humanistic Mathematics, *Alvin White,* Editor.

33. Research Issues in Undergraduate Mathematics Learning: Preliminary Analyses and Results, *James J. Kaput and Ed Dubinsky,* Editors.

34. In Eves' Circles, *Joby Milo Anthony,* Editor.

35. You're the Professor, What Next? Ideas and Resources for Preparing College Teachers, *The Committee on Preparation for College Teaching, Bettye Anne Case,* Editor.

36. Preparing for a New Calculus: Conference Proceedings, *Anita E. Solow,* Editor.

37. A Practical Guide to Cooperative Learning in Collegiate Mathematics, *Nancy L. Hagelgans, Barbara E. Reynolds, SDS, Keith Schwingendorf, Draga Vidakovic, Ed Dubinsky, Mazen Shahin, G. Joseph Wimbish, Jr.*

38. Models That Work: Case Studies in Effective Undergraduate Mathematics Programs, *Alan C. Tucker,* Editor.

39. Calculus: The Dynamics of Change, *CUPM Subcommittee on Calculus Reform and the First Two Years, A. Wayne Roberts,* Editor.

40. Vita Mathematica: Historical Research and Integration with Teaching, *Ronald Calinger,* Editor.

41. Geometry Turned On: Dynamic Software in Learning, Teaching, and Research, *James R. King and Doris Schattschneider,* Editors.

42. Resources for Teaching Linear Algebra, *David Carlson, Charles R. Johnson, David C. Lay, A. Duane Porter, Ann E. Watkins, William Watkins,* Editors.

43. Student Assessment in Calculus: A Report of the NSF Working Group on Assessment in Calculus, *Alan Schoenfeld,* Editor.

44. Readings in Cooperative Learning for Undergraduate Mathematics, *Ed Dubinsky, David Mathews, and Barbara E. Reynolds,* Editors.

MAA Service Center
P. O. Box 91112
Washington, DC 20090-1112
800-331-1622   fax: 301-206-9789

Contents

Foreword

In 1980 Ronald Reagan was elected president of the United States, IBM began an urgent program to develop a personal computer for which Microsoft agreed to provide the operating system, and Benoit Mandelbrot got his first real look at the archetypal fractal, the eponymous Mandelbrot set. One hundred years from now, which of these events is more likely to be remembered as having had the greatest influence on science and human affairs?

At this moment, two decades later, personal computers seem to be well in the lead. PCs are on every desk, all now connected by the world-linking Internet. The personal computer has truly transformed the way the world works. No invention since the printing press has created such a widespread impact and no human activity has ever changed society so quickly.

However, when we examine Internet patterns in detail, guess what we find just beneath the surface? The footprints of fractals. These wondrous geometric objects, discovered by Mandelbrot just over twenty years ago, turn out to be the key to understanding the frenetic behavior of signals linking the world's computers, as well as the means to efficient compression that makes it possible to transmit images over the Internet. Without fractals, engineers would never have been able to make the World Wide Web work as well as it does. And without the Web, PCs would be just one more labor-saving appliance.

Computers are important instrumentally; they provide tools that enable us to work more efficiently, to see patterns previously hidden, and to organize information in new and revealing ways. In contrast, fractals are important fundamentally; they provide elements of a totally new geometry that offer a profoundly different way to understand nature. In the long run, this new understanding of nature will count for far more than momentary advances in technology or politics.

The predominant Western view of the relation between mathematics and nature is a legacy of Plato's distinction between a world of ideals and a world of actualities. Mathematics, in this view, belongs to the ethereal world of ideals; nature, being earthly rather than heavenly, belongs to a world of actualities that is both imperfect and incomplete. Thus, reality is best understood by approximation in terms of ideal mathematical models.

Our most important inheritance from this tradition is Euclidean geometry, the axiomatic study of lines, circles, and triangles that form an ideal (and therefore approximate) basis for understanding geography, mechanics, astronomy, and everything real. From this perspective, nature is like noisy mathematics—rough and crumpled, slightly out of focus.

Fractals create an alternative to Euclidean geometry whose elements are not lines and circles but iterations and self-similarities, whose surfaces are not smooth but jagged, whose features are not perfect but broken. Derived from apparent pathologies that puzzled or affronted traditional mathematicians, fractals reveal an entirely new geometry that enables us to understand formerly inexplicable real-world phenomena.

Fractals provide insight into the distribution of galaxies, the spread of bacterial colonies, the grammar of DNA, the shape of coastlines, changes in climate, development of hurricanes, growth of crystals, percolation of ground pollutants, turbulence of fluids, and the path of lightning. They have been employed to create powerful antennas, to develop fiber optics, to monitor financial data, to compress images, and to produce artificial landscapes. Their influence has been felt in art, architecture, drama (*Arcadia*), film (*Jurassic Park*), music, and poetry. Fractals have even penetrated the inner sanctum of elite culture—New Yorker cartoons.

For many, this extraordinary utility would be a sufficient warrant for fractals to be awarded a prominent role in mathematics education. But reasons other than utility can also be advanced, and it is these reasons, not utility, that form the major thrust of this volume.

Simply put, fractals enable everyone to enjoy mathematics. Nothing else can make such a striking—and important—claim.

Arguments about strategies for improving mathematics education roil state and local school politics. Some urge strict exam-enforced standards; others advocate more inviting contexts for learning. Some promote traditional curricula; others support enhanced or integrated programs. Rarely do the protagonists in these "math wars" stop to ask whether different mathematics might yield increased learning.

But that is precisely the argument advanced by the authors represented in this volume. They focus on teaching fractals, not primarily because fractals are important but because learning about fractals is, as one student put it, "indescribably exciting... and uniquely intriguing." It is easy to see why:

- The first steps are so much fun. Exploring fractals creates unprecedented enthusiasm for discovery learning among both teachers and students.

- Fractals are beautiful. Stunning visuals appeal to the mind's eye and create contagious demand for continued exploration.

- Anyone can play. Exploration of fractal geometry appeals to students of every age, from primary school through college and beyond.

- Fractals promote curiosity. Simple rules, easily modified, create nearly uncontrollable temptations to explore different options to see what surprising patterns will emerge.

- Simple ideas lead to unexpected complexity. Fractals are more life-like than objects studied in other parts of mathematics; thus they appeal to many students who find traditional mathematics cold and austere.

- Many easy problems remain unsolved. Fractals are rich in open conjectures that lead to deep mathematics. Moreover, the distance from elementary steps to unsolved problems is very short.

- Careful inspection yields immediate rewards. Insight and conjectures arise readily when our well-developed visual intuition is applied to fractal images. In studying fractals, children can see and conjecture as well as adults.

- Computers enhance learning. The visual impact of computer graphics makes fractal images unforgettable, while the unforgiving logical demands of computer programs yield important lessons in the value of rigorous thinking.

The history of mathematics education is long and convoluted, reflecting both the changing nature of mathematics and the evolving demands of society. Although in the eighteenth and nineteenth centuries mathematics was both experimental and theoretical, during much of the twentieth century the theoretical aspect has dominated. Much of mathematics education followed this trend towards theory and abstraction, leading to alarming reports of rising mathematical illiteracy not only in the United States but in many other countries as well.

Fractals represent a rebirth of experimental mathematics, enabled by computers and enhanced by powerful evidence of utility. In the ebb and flow of mathematical fashion, the struggle between theoretical and experimental is once again more nearly in balance. What remains is the challenge of restoring this balance to mathematics education. It is to that important task that this book is devoted.

Lynn Arthur Steen
St. Olaf College
Northfield, MN
April 13, 2000.

Preface

When revising this preface for the last time, we suddenly realized that the terms "graphics" and "mathematics education" in the title are defined nowhere, but this will not upset the reader. More seriously, "fractals" are not defined either. This question has bedeviled the field since Mandelbrot (1975) and is dealt with in an *Overview* chapter of Mandelbrot (2002), which deserves to be summarized here.

There is no satisfactory formal definition. Informally, fractal geometry is the systematic study of certain very irregular shapes, in either mathematics or the real world, that are scale invariant. This means that each small part is very much like a reduced copy of the whole. This wording suggests a familiar term that occurs rarely in this book, but underlies many explicit aspects of fractal geometry and is close to providing the best one-word characterization of its scope. This word is *roughness*: fractal geometry is the first-ever scientific approach to our sensation of rough versus smooth.

Many sciences arose directly from the "five" senses. Visual sensations led to the notions of bulk and shape and of brightness and color. The sensation of heavy versus light led to mechanics and the sensation of hot versus cold led to the theory of heat. Each of those sciences began with a special case that is technically manageable, widespread and important. In the theory of sound, that special case was not provided by chirps or drums, but by string instruments, whose sinusoidal vibrations are translation invariant. Against this background, the sensation of rough versus smooth has been extraordinarily slow in becoming formalized. The first approximation on which a broader theory could lean was provided by fractals, which are not translation, but dilation invariant. Consequently, we can both characterize fractal geometry, and motivate its ubiquity, by describing it as the first stage of the study of roughness.

***

Be that as it may, fractal geometry has many other aspects. What is of concern for this book is that for over a decade it has had an increasingly strong impact on mathematics and science education. This influence is unusual in both range (elementary to graduate school) and ubiquity (mathematics, computer science, seemingly every natural science and engineering discipline, but also economics, art and architecture, music, literature, history, and education). Much of this development came from dedicated teachers working independently, each emphasizing those points he or she felt were the most effective in the local setting. In our long list of scientific meetings on fractal geometry we find only one devoted entirely to the teaching of fractals. On December 12 and 13, 1997, we ran such a meeting at Yale University under the sponsorship of the National Science Foundation. This volume's title is borrowed from the address one of us gave at ICME-7 in 1992. The text brings together contributions by most who attended the meeting, and a few who were invited but could not attend because of schedule conflicts. In editing, we preserved the individual voices, feeling their differences echo the great variety of approaches to teaching fractal geometry. This is still a very personal enterprise, but as we shall see, some common themes have arisen.

While developments outside mathematics are of central importance for convincing our students of the relevance of fractal geometry—the beauty is self-evident—contributions were solicited from teachers well-versed in the technical details of the subject. In addition, participants were selected (mostly) geographically, and also to insure representation of a wide range of institutions: public and private high schools, small liberal arts colleges, large research universities, and a community college.

Special mention must be made of the fractal geometry program at Florida Atlantic University. (See Chapter 9.) FAU is in one of the most rapidly growing areas of the country; its students experience the associated problems. Because of its success, this program has expanded to all of Broward and Dade counties in Florida. We take this as a very good sign.

Despite this considerable range of institutions and levels, discussions revealed much that was common. Most participants described their courses and focused on one or two particular features, ranging from student projects to the use of web sites and Java applets in teaching fractals. More than just some core topics, we found common themes in the success of these courses. Many students are fascinated by the beautiful graphics, but also by the experimental nature of the subject. Even at the level of fourth grade children, students can master enough of the geometry of plane transformations to construct fractal images. The computer is central in this— being able to change a parameter and immediately see the often surprising effect on the resulting fractal, is certainly one of the most successful ways to encourage students to strike out on their own. Discovery learning is central to fractal geometry.

Another common feature to emerge is that enthusiasm for the subject forms a link between students and teachers. In standard courses, the material is old and familiar to us as teachers. In matters of content, often we teach these subjects as we learned them. But fractal geometry is new, was not yet born when most of us were students. The field is growing so rapidly that keeping up with everything is almost impossible. Using the web, our students may find new developments with which we are unfamiliar. The newness of fractal geometry deconstructs some of the differences between teachers and students, makes the classroom more egalitarian, and allows us to approach the subject with genuine child-like wonder. This enthusiasm feeds both students and teachers.

Three aspects of fractal geometry contribute to its continued success in the classroom. First is the very short distance between elementary constructions and problems unsolved despite energetic efforts by many skilled mathematicians. Second, not all the easy results are known yet. Third, new results and directions are arising all the time. Examples of all three aspects were discussed at the meeting. Concrete suggestions of how some of us approached problems others have encountered also figured in the discussion.

We plan additional meetings, of wider scope, and are well-aware that this volume barely scratches the surface of its topic.

We close this preface with comments about two courses at Yale. One is a general education mathematics course that grew from projects envisioned by Benoit Mandelbrot and reflected by an address he delivered in 1992 (Chapter 3), and from experience acquired by Michael Frame at Union College. *Fractal Geometry for Non-Science Students* (Math 190) was first offered in the spring of 1993 and attracted a huge following, with over 160 students, mostly from the humanities and social sciences. (Yale's total undergraduate population numbers about 5300.) First taught by Michael Frame, this course has run in several formats and varying enrollments under Richard Voss, Frank Geschwin, and Ken Monks. Most recently (fall of 1999 and of 2000) it was taught by Michael Frame, with over 170 students both semesters. We believe it is important for courses such as this to have homes in Mathematics Departments. This emphasizes the impact of fractal geometry on the well-known problems of the perceived difficulty of and interest in mathematics courses.

All the lecture material of the fall 2000 offering of *Fractal Geometry for Non-Science Students* was presented as webpages, accessible from students' personal computers. The heavy reliance on visuals made this approach natural, and freed the students from the burden of taking notes. The course ran at a faster pace, with a slight (perhaps statistically insignificant) increase in grades, and covered more material than in previous years. We believe fractal geometry is especially well-suited for web-based instruction. The current version of the webpages is available at

```
http://www.maa.org/Fractals/Welcome.html
```

Encouraged by this success, we are investigating a distance learning version of this course.

Since the fall of 1995, a parallel higher-level course, *Fractal Geometry: Techniques and Applications* (Math 290) has been offered for science and engineering undergraduates. We decided not to write about these courses separately, in part because this volume already is quite large, and also because lower level course's success is due largely to its emphasis on applications *outside* of science. These applications are surveyed in the *Panorama of Fractals and their Uses*, a web document growing at

```
http://www.maa.org/Fractals/Panorama/
                Welcome.html
```

See also the *Appendix*. Additional comments about both courses are woven throughout Chapters 1–3. Nevertheless, it was the remarkable success of *Fractal Geometry for Non-Science Students* that fueled much of our enthusiasm for the educational value of fractals. In a real sense, that course is responsible for this volume.

A note about references

References are in a single combined list at the end of this volume. The prefixes s, v, and w refer to the software, video, and web references near the end of the list.

Acknowledgements

The first offering of Math 190 was expedited by the support of Richard Beals and Roger Howe of the Yale Mathematics Department, and by Richard Brodhead, Dean of Yale College. Over the years, the Mathematics Department has been a comfortable home for both Math 190 and 290. In the final preparation of this book we benefitted greatly from the assistance of Jean Maatta and Nial Neger. Nial proofread the entire text and corrected several serious pedagogical errors by the editors. Jean prepared the references from 16 separate lists, and helped with the proofreading. Comments by the several reviewers improved the readability of the text. The editors benefited from familiarity with the fractal geometry summer workshop program run by Denny Gulick of the University of Maryland and Jon Scott of Montgomery College. Through grants DUE 9455636 and DMS 9983403, the National Science Foundation provided support for projects integral to the writing of this book. Finally, the editors are pleased to acknowledge the patience of the contributors in waiting for this volume to appear.

Michael Frame and Benoit B. Mandelbrot

New Haven, CT
August, 2001

Introductory Essays

Chapter 1

Some Reasons for the Effectiveness of Fractals in Mathematics Education

Benoit B. Mandelbrot and Michael Frame

Short is the distance between the elementary and the most sophisticated results, which brings rank beginners close to certain current concerns of the specialists. There is a host of simple observations that everyone can appreciate and believe to be true, but not even the greatest experts can prove or disprove. There is a supply of unsolved, elementary problems that give students the opportunity to learn how mathematics can be done by enabling them to do new (if not necessarily earth-shaking) mathematics; there is a continuing flow of new results in unexpected directions.

1 Introduction

In the immediate wake of Mandelbrot (1982), fractals began appearing in mathematics and science courses, mostly at the college level, and usually in courses on topics in geometry, physics, or computer science. Student reaction often was extremely positive, and soon entire courses on fractal geometry (and the related discipline of chaotic dynamics) arose. Most of the initial offerings were aimed at students in science and engineering, and occasionally economics, but, something about fractal geometry resonated for a wider audience. The subject made its way into the general education mathematics and science curriculum, and into parts of the high school curriculum. Eventually, entire courses based on fractal geometry were developed for humanities and social sciences students, some fully satisfy the mathematics or science requirement for these students. As an introduction to this volume, we share some experiences and thoughts about the effectiveness and appropriateness of these courses.

As teachers, we tell our students to first present their case and allow the objections to be raised later by the devil's advocate. But we decided to preempt some of the advocate's doubts or objections before we move on with our story.

1.1 The early days

A few years ago, the popularity of elementary courses using fractals was largely credited to the surprising beauty of fractal pictures and the centrality of the computer to instruction in what lies behind those pictures. A math or science course filled with striking, unfamiliar visual images, where the computer was used almost every day, sometimes by the students? The early general education fractals courses did not fit into the standard science or mathematics format, a novel feature that contributed to their popularity.

1.2 What beyond novelty?

We shall argue that novelty was neither the only, nor the most significant factor. But even if it had been, and if the popularity of these courses had declined as the novelty wore off, so what? For a few years we would have had effective vehicles for showing a wide audience that science is an ongoing process, an exciting activity pursued by living people. While introductory courses for majors are appropriate for some non-science students, and qualitative survey courses are appropriate for some others, fractal geometry provided a middle ground between quantitative work aiming toward some later reward (only briefly glimpsed by students not going beyond the introductory course), and qualitative, sometimes journalistic, sketches. In general education fractal geometry courses, students with only moderate skills in high school algebra could learn to do certain things themselves rather than read forever about what others had done. They could grow fractal trees, understand the construction of the Mandelbrot and Julia sets, and synthesize their own fractal mountains and clouds. Much of this mathematics spoke directly to their visible world. Many came away from these courses feeling they had understood some little bit of how the world works. And the very fact that some of the basic definitions are unsettled, and that there are differences of opinion among leading players, underscored the human aspect of science. No longer a crystalline image of pure deductive perfection, mathematics is revealed to be an enterprise as full of guesses, mistakes, and luck as any other creative activity. Even if the worst fears had been fulfilled, we would have given several years of humanities and social science students a friendlier view of science and mathematics.

Fortunately, anecdotal evidence suggests that, while much of the standard material and computerized instruction techniques are no longer novel, the audience for fractal geometry courses is not disappearing, thus disproving those fears.

1.3   What aspects of novelty have vanished?

Success destroyed part of the novelty of these courses. Now images of the Mandelbrot set appear on screen savers, T-shirts, notebooks, refrigerator magnets, the covers of books (including novels), MTV, basketball cards, and as at least one crop circle in the fields near Cambridge, UK. Fractals have appeared in novels by John Updike, Kate Wilhelm, Richard Powers, Arthur C. Clarke, Michael Crichton, and others. Fractals and chaos were central to Tom Stoppard's play *Arcadia*, which includes near quotes from Mandelbrot. Commercial television ("Murphy Brown," "The Simpsons," "The X-Files"), movies ("Jurassic Park"), and even public radio ("A Prairie Home Companion") have incorporated fractals and chaos. In the middle 1980s, fractal pictures produced "oohhs," "aahhs," and even stunned silence; now they are an ingrained part of both popular and highbrow culture (the music of Wuorinen and Ligeti, for example). While still beautiful, they are no longer novel.

A similar statement can be made about methodology. In the middle 1980s, the use of computers in the classroom was uncommon, and added to the appeal of fractal geometry courses. Students often lead faculty in recognizing and embracing important new technologies. The presence of computers was a definite draw for fractal geometry courses. Today, a randomly selected calculus class is reasonably likely to include some aspect of symbolic or graphical computation, and many introductory science classes use computers, at least in the lab sections. The use of computers in many other science and mathematics courses no longer distinguishes fractal geometry from many other subjects.

1.4   Yet these courses' popularity survived their novelty. Why is this?

Instead of being a short-lived fad, fractal geometry survived handsomely and became a style, part of our culture.

The absence of competition is one obvious reason: fractal geometry remains the most visual subject in mathematics and science. Students are increasingly accustomed to thinking pictorially (witness the stunning success of graphical user interfaces over sequences of command lines) and continue to be comfortable with the reasoning in fractal geometry. Then, too, in addition to microscopically small and astronomically large fractals, there is also an abundance of human-sized fractals, whereas there are not human-sized quarks or galaxies.

Next, we must mention surprises. Students are amazed the first time they see that for a given set of rules, the deterministic IFS algorithm produces the same fractal regardless of the starting shape. The gasket rules make a gasket from a square, a single point, a picture of your brother,... anything. If the Mandelbrot set is introduced by watching videotapes of animated zooms, then the utter simplicity of the algorithm generating the Mandelbrot set is amazing. Part of what keeps the course interesting is the surprises waiting around almost every corner. Also, besides science and mathematics, fractals have direct applications in many fields, including music, literature, visual art, architecture, sculpture, dance, technology, business, finance, economics, psychology, and sociology. In this way, fractals act as a sort of common language, *lingua franca*, allowing students with diverse backgrounds to bring these methods into their own worlds, and in the context of this language, better understand some aspects of their classmates' work.

Three other reasons are more central to the continued success of general education fractal geometry courses. By exploiting these reasons, we keep strengthening current courses and finding directions for future development.

As a preliminary, let us briefly list these reasons for the pedagogical success of fractal geometry. We shall return to each in detail.

1.4.1   First, a short distance from the downright elementary to the hopelessly unsolved

First surprise: truly elementary aspects of fractal geometry have been successfully explained to elementary school students, as seen in Chapters 10 and 13. From those aspects, there is an uncannily short distance to unsolved problems. Few other disciplines—knot theory is an example—can make this claim.

Many students feel that mathematics is an old, dead subject. And why not? Most of high school mathematics was perfected many centuries ago by the Greeks and Arabs, or at the latest, a few centuries ago by Newton and Leibnitz. Mathematics appears as a closed, finished subject. To counter that view, nothing goes quite so far as being able to understand, after only a few hours of background, problems that remain unsolved today. Number theory had a standard unsolved but accessible problem that need not be named. Alas, that problem now is solved. Increasing our emphasis on unsolved problems brings students closer to an edge of our lively, growing field and gives them some real appreciation of science and mathematics as ongoing processes.

1.4.2 Second, easy results remain reachable

The unsolved problems to which we alluded above are very difficult, and have been studied for years by experts. In contrast, not nearly all the easy aspects of fractal geometry have been explored. At first, this may seem more relevant to graduate students, but in fact, plenty of the problems are accessible to bright undergraduates. The National Conference of Undergraduate Research and the Hudson River Undergraduate Mathematics Conference, among others, include presentations of student work on fractal geometry. It may be uncommon for students in a general education course to make new contributions to fractal geometry (though to be sure they often come up with very creative projects applying fractal concepts to their own fields), but their classmates in sciences and mathematics can and do. (See Frame & Lanski (1999).) Incorporating new work done by known, fellow undergraduates can have an electrifying effect on the class. Few things bring home the accessibility of a field so much as seeing and understanding something new done by someone about the same age as the students. Then, too, this is quite exciting for the science and mathematics students whose work is being described. And it can be, and has been, a catalyst for communication between science and non-science students. So far as we know, in no other area of science or mathematics are undergraduates so likely to achieve a sense of ownership of material.

1.4.3 Third, new topics continue to arise and many are accessible

New things, accessible at some honest level, keep arising in fractal geometry. Of course, new things are happening all around, but the latest advances in superstring theory, for example, cannot be described in any but the most superficial level in a general education science course. This is not to say all aspects of fractal geometry are accessible to nonspecialists. Holomorphic surgery, for instance, lives in a pretty rarefied atmosphere. And there is deep mathematics underlying much of fractal geometry. But pictures were central to the birth of the field, and most open problems remain rooted in visual conjectures that can be explained and understood at a reasonable level without the details of the supporting mathematics. While undergraduates can do new work, it is unlikely to be deep work. In fractal geometry much of even the current challenging new work can be presented only in part but, honestly, and without condescension to our students.

Later we shall further explore some aspects of each of these points.

1.5 Most important of all: curiosity

Teaching endless sections of calculus, precalculus, or baby statistics to uninterested audiences is hard work and all too often we yield to the temptation to play to the lowest third of the class. The students merely try to survive their mathematics requirement. Little surprise we complain about our students' lack of interest, and about the disappearance of childlike curiosity and sense of wonder.

Fractal geometry offers an escape from this problem. It is risky and doesn't always work, for it relies on keeping this youthful curiosity alive, or reawakening it if necessary. In the final *Calvin and Hobbes* comic strip, Calvin and Hobbes are on a sled zipping down a snow-covered hill. Calvin's final words are, "It's a magical world, Hobbes ol' buddy. Let's go exploring!" This is the feeling we want to awaken, to share with our students.

Teaching in this way, especially emphasizing the points we suggest, demands faith in our students. Faith that by showing them unsolved problems, work done by other students, and new work done by scientists, they will respond by accepting these offerings and becoming engaged in the subject. It does not always work. But when it does, we have succeeded in helping another student become a more scientifically literate citizen. Surely, this is a worthwhile goal.

2 Instant gratification: from the elementary to the diabolic and unsolved, the shortest distance is . . .

In most areas of mathematics, or indeed of science, a vast chasm separates the beginner from even understanding a statement of an unsolved problem. The Poincarè conjecture is a very long way from a first glimpse of topological spaces and homotopies. Science and mathematics courses for non-majors usually address unsolved problems in one of two ways: complete neglect or vast oversimplification. This can leave students with the impression that nothing remains to be done, or that the frontiers are far too distant to be seen; neither picture is especially inviting.

Fractal geometry is completely different. While the *solutions* of hard problems often involve very clever use of sophisticated mathematics, frequently the *statements* do not. Here we mention two examples, to be amplified and expanded on in the next chapter.

The first observed example of Brownian motion occurred in a drop of water: pollen grains dancing under the impact of molecular bombardment. Nowadays this can be demonstrated in class with rather modest equipment: a microscope fitted with a video camera and a projector. Increasing the magnification reveals ever finer detail in the dance, thus providing a visual hint of self-similarity. A brief description of Gaussian distributions—or even of random walk—is all we need to motivate computer simulations of Brownian motion. Taking a Brownian path for a finite duration and subtracting the linear interpolation from the initial point to the final point produces a Brownian plane cluster. The periphery, or *hull*, of this cluster looks like the coastline of an island. Together with numerical experiments, this led to the conjecture that the hull has dimension 4/3. Dimension is introduced early in fractal geometry classes, so freshman English majors can understand this conjecture. Yet it is unproved.[1]

No icon of fractal geometry is more familiar than the Mandelbrot set. Its strange beauty entrances amateurs and experts alike. Many credit it with the resurgence of interest in complex iteration theory, and its role in the birth of computer-aided experimental mathematics is incalculable. For students, the first surprise is the simplicity of the algorithm to generate it. For each complex number c, start with $z_0 = 0$ and produce the sequence $z_1, z_2, \ldots$ by $z_{i+1} = z_i^2 + c$. The point c belongs to the Mandelbrot set if and only if the sequence remains bounded. How can such a simple process make such an amazing picture? Moreover, a picture that upon magnification reveals an infinite variety of patterns repeating but with variations. One way for the sequence to remain bounded is to converge to some repeating pattern, or cycle. If all points near to $z_0 = 0$ produce sequences converging to the same cycle, the cycle is stable. Careful observation of computer experiments led Mandelbrot to conjecture that arbitrarily close to every point of the Mandelbrot set lies a c for which there is a stable cycle. All of these concepts are covered in detail in introductory courses, so here, too, beginning students can get an honest understanding of this conjecture, unsolved despite heroic effort.

3 Some easy results remain: "There's treasure everywhere"

3.1 Discovery learning

Learning is about discovery, but undergraduates usually learn about past discoveries from which all roughness has been polished away giving rise to elegant approaches. Good teaching style, but also speed and efficiency, lead us to present mathematics in this fashion. The students' act of discovery dissolves in becoming comfortable with things already known to us. Regardless of how gently we listen, this is an asymmetric relationship: we have the sought-after knowledge. We are the masters, the final arbiters, they the apprentices.

In most instances this relationship is appropriate, unavoidable. If every student learned mathematics and science by reconstructing them from the ground up, few would ever see the wonders we now treasure. Which undergraduate would have discovered special relativity? But for most undergraduate mathematics and science students, and nearly all non-science students, this master-apprentice relationship persists through their careers, leaving no idea of how mathematics and

---

[1] Stop the presses: this conjecture has been proved in Lawler, Werner, & Schramm (2000).

science are done. Fractal geometry offers a different possibility.

Term projects are a central part of our courses for both non-science and science students. To be sure, some projects turn out less appropriate than hoped, but many have been quite creative. Refer to the *student project* entries in *A Guide to the Topics*. Generally, giving a student an open-ended project and the responsibility for formulating at least some of the questions, and being interested in what the student has to say about these questions, is a wonderful way to extract hard work.

3.2 A term project example: connectivity of gasket relatives

We give one example, Kern (1997), a project of a freshman in a recent class. Students often see the right Sierpinski gasket as one of the first examples of a mathematical fractal. The IFS formulation is especially simple: this gasket is the only compact subset of the plane left invariant by the transformations

$$T_1(x, y) = \left(\frac{x}{2}, \frac{y}{2}\right),$$

$$T_2(x, y) = \left(\frac{x}{2}, \frac{y}{2}\right) + \left(\frac{1}{2}, 0\right),$$

$$T_3(x, y) = \left(\frac{x}{2}, \frac{y}{2}\right) + \left(0, \frac{1}{2}\right).$$

Applying these transformations to the unit square $S = \{(x, y) : 0 \leq x \leq 1, 0 \leq y \leq 1\}$ gives three squares $S_i = T_i(S)$ for $i = 1, 2, 3$. Among the infinitely many changes of the T_i, in general producing different fractals, a particularly interesting and manageable class consists of including reflections across the x- and y-axes, rotations by $\frac{\pi}{2}$, π, and $\frac{3\pi}{2}$, and appropriate translations so the three resulting squares occupy the same positions as $T_1(S)$, $T_2(S)$, and $T_3(S)$. Pictures of the resulting fractals are given on pgs 246–8 of Peitgen, Jurgens & Saupe (1992a).

What sort of order can be brought to this table of pictures? Connectivity properties may be the most obvious: they allow one to classify fractals.

dusts (totally disconnected, Cantor sets),

dendrites (singly connected throughout, without loops),

multiply connected (connected with loops), and

hybrids (infinitely many components each containing a curve).

A parameter space map, painting points according to which of the four behaviors the corresponding fractal exhibits, did not reveal any illuminating patterns. However, sometimes (though not always—certainly not in the Cantor set cases, for example) in the unit square S there are finite collections of line segments that are preserved in $T_1(S) \cup T_2(S) \cup T_3(S)$. In

Figure 1: Relatives of the Sierpinski gasket: Cantor dust, dendrite, multiply connected, and hybrid. Can you find preserved line segments in the last three?

the cases where they could be found, these did give a transparent reason for the connectivity properties. This approach was generated by the student, looking for patterns by staring at the examples for hours on end.

What can we make of the observation that different collections of line segments work for different IFS? The student speculated that there is a *universal shape*, perhaps a union of some of the line segments from several examples, whose behavior under one application of T_1, T_2, and T_3 determines the connectivity form of the limiting fractal. This is an excellent question to be raised by a freshman, especially in a self-directed investigation.

This is just one example. Fractal geometry may be unique in providing such a wealth of visually motivated, but analytically expressed, problems. Truly, there is treasure everywhere.

4   Something new is always happening

New mathematics is coming up all the time; ours is a very lively field. However, many new developments are at an advanced level, often comprehensible only to experts having years of specialized training. To be sure, deep mathematical discoveries abound in fractal geometry, too. But because pictures are so central, here many advances have visual expressions that honestly reveal some of the underlying mathematics. New developments in retroviruses or in quantum gravity are unlikely to be comprehensible at anything other than a superficial level to general education students. They hear *about* the advances, but not *why* or *how* they work. The highly visual aspect of fractal geometry has allowed us to incorporate the most recent work into our courses in a serious way.

Here we describe one new development, and mention another to be explored in the next chapter.

4.1   Fractal lacunarity

It is difficult to imagine an introductory course on fractals that does not include computing dimensions of self-similar

fractals. (See Chapters 5, 12, and 15, for example.) The calculations are straightforward, a skill mastered without excessive effort. Moreover, the idea generalizes to data from experiments, opening the way for a variety of student projects. However, one of the earliest exercises we assign points out a limitation of dimension: quite different-looking sets can have the same dimension. For example, all four fractals in Figure 1 have dimension $\log(3)/\log(2)$. The Sierpinski carpets of Figure 2 (Plate 318 of Mandelbrot (1982)) both subdivide the unit square into 49 pieces, each scaled by $\frac{1}{7}$, and delete nine of these pieces. So both have dimension $\log(40)/\log(7)$. On the left, these holes are distributed uniformly, on the right they are clustered together into one large hole in the middle. *Lacunarity* is one expression of this difference, and is another step in characterizing fractals through associated numbers. Here the number represents the distribution of holes or gaps, *lacunae*, in the fractal. This reinforces for students the relation between numbers and the visual aspects they are meant to represent. But also, this is current work, and even some of the basic issues are not yet settled. With this, our students see science as it is developing, and can understand some components of the debate.

To give an example of the kinds of results accessible to students having some familiarity with sequences and calculus, we describe an approach to the fractals of Figure 2. For a subset $A \subset \mathbf{R}^2$, the ϵ-thickening is defined as

$$A_\epsilon = \{\mathbf{x} \in \mathbf{R}^2 : d(\mathbf{x}, \mathbf{y}) \leq \epsilon \text{ for some } \mathbf{y} \in A\}$$

where $d(\mathbf{x}, \mathbf{y})$ is the Euclidean distance between $\mathbf{x}$ and $\mathbf{y}$.

Now suppose A is either of the Sierpinski carpets in Figure 2. For large ϵ, A_ϵ fills all the holes of A and the area of A_ϵ, $|A_\epsilon|$, is $1 + 4\epsilon + \pi\epsilon^2$. As $\epsilon \to 0$, the holes of A become visible and increase the rate at which $|A_\epsilon|$ decreases. Calculations with Euclidean shapes—points, line segments, and circles, for example—show $|A_\epsilon| \approx L \cdot \epsilon^{2-d}$, where d is the dimension of the object. This relation can be used to compute the dimension, a technique developed by Minkowski and Bouligand. A first approach to lacunarity is the prefactor L, or more precisely, $1/L$, if the limit exists.

Figure 2: Two Sierpinski carpet fractals with the same dimension.

A general Sierpinski carpet is made with initiator the filled-in unit square, and generator the square with M squares of side length s removed. The iteration process next covers the complement of these M holes with N copies of the generator, each scaled by r. (Note the relation $1 - Ms^2 = Nr^2$.) For the carpet on the left side of Fig. 2 we see $M = 9$, $s = \frac{1}{7}$, $N = 40$, and $r = \frac{1}{7}$; on the right $M = 1$, $s = \frac{3}{7}$, $N = 40$, and $r = \frac{1}{7}$.

It is well known that for the box-counting dimension the limit as $\epsilon \to 0$ can be replaced by the sequential limit $\epsilon_n \to 0$, for ϵ_n satisfying mild conditions. Although the prefactor is generally more sensitive than the exponent, we begin with the sequence $\epsilon_n = sr^{n-1}/2$. For Sierpinski carpets A it is not difficult to see A_{ϵ_n} fills all holes of generation $\geq n$, while holes of generation $m < n$ remain. They are squares of side length $s(r^{m-1} - r^{n-1})$. Straightforward calculation gives

$$|A_{\epsilon_n}| = (4\epsilon_n + \pi\epsilon_n^2) + Ms^2\left(\left(\frac{2}{1 - Nr}r^n - \frac{1}{1 - N}r^{2n}\right) + (Nr^2)^n\left(\frac{1}{1 - Nr^2} - \frac{2}{1 - Nr} + \frac{1}{1 - N}\right)\right).$$

Using $L \approx |A_{\epsilon_n}|\epsilon_n^{d-2}$, we obtain

$$L \approx M2^{2-d}s^d\left(\frac{1}{1 - Nr^2} - \frac{2}{1 - Nr} + \frac{1}{1 - N}\right).$$

Substituting in the values of M, s, N, and r, we obtain $L \approx 1.41325$ and $L \approx 1.26026$ for the left and right carpets. So provisionally, the lacunarities are 0.707589 and 0.793487, agreeing with the notion that higher lacunarity corresponds to a more uneven distribution of holes.

Unfortunately, different sequences ϵ_n can give different values of L. Several approaches are possible, but one that is relatively easy to motivate and implement is to use a logarithmic average

$$\lim_{T \to \infty} \frac{1}{T} \int_0^T \frac{|A_{e^{-t}}|}{(2e^{-t})^{2-d}} dt.$$

The 2 in the denominator is a normalizing factor. For these carpets, this reduces to

$$\frac{Ms^d}{\log(1/r)}\left(\frac{1}{1 - Nr^2}\frac{1 - r^{2-d}}{2 - d} - \frac{2}{1 - Nr}\frac{1 - r^{1-d}}{1 - d} + \frac{1}{1 - N}\frac{1 - r^{-d}}{-d}\right).$$

Substituting in the values of M, s, N, and r, we obtain 1.305884 and 1.164514 for the left and right carpets. The respective lacunarities are 0.765765 and 0.858727.

These calculations involve simple geometry and can be extended easily to gaskets, their relatives, and the like. Even as the concepts continue to evolve, this is a rich source of ideas for student projects. Comparison with other lacunarity candidate measures—crosscut (Mandelbrot, Vespignani & Kaufman (1995)) and antipodal correlations (Mandelbrot & Stauffer (1994)), among others—in simple cases, is yet another source of projects. This has proven especially interesting because it shows students first-hand some of the issues involved in defining a measurement of a delicate property. Without being too heavy-handed, we point out in calculus that the definitions have been well-established for centuries. And even students in general education courses can appreciate the visual issues involved in the clustering of the *lacunae*.

4.2 Fractals in finance

As of this writing, the most common models of the stock market are based on Brownian motion. In fact, the first mathematical formulation of Brownian motion was Louis Bachelier's 1900 model of the Paris bond market. However, comparison with data instantly reveals many unrealistic features of Brownian motion $X(t)$. For example, $X(t_1) - X(t_2)$ and $X(t_3) - X(t_4)$ are independent for $t_1 < t_2 < t_3 < t_4$, and $X(t_1) - X(t_2)$ is Gaussian distributed with mean 0 and variance $|t_1 - t_2|$. That is, increments over disjoint time intervals are independent of one another, and the increments follow the familiar bell curve, so large increments are very rare. The latter is called the *short tails* property.

Are these reasonable features of real markets? Why should price changes one day be independent of price changes on a previous day? Moreover, computing the variance from market data assembled over a very long time, events of 10σ, for example, occur with enormously much higher frequency than the Gaussian value, which is (!) 10^{-24}. Practitioners circumvent these problems by a number of *ad hoc* fixes, adding up to a feeling similar to that produced by Ptolemy's cosmology: add enough epicycles and you can match any observed motion of the planets. Never mind the problems produced by the physicality of the epicycles, among other things. (Of course, in finance the situation is much worse. No one has a collection of epicycles that predicts market behavior with any reliability at all.)

In the 1960s, Mandelbrot proposed two alternatives to Brownian motion models. Mandelbrot (1963) had increments governed by the Lévy stable distribution (so with long tails), but still independent of one another. In 1965 Mandelbrot proposed a model based on fractional Brownian motion (See Mandelbrot (1997).) This model consequently had increments that are dependent, though still governed by the Gaussian distribution. Both are improvements, in different ways, of the Brownian motion models.

It is a considerable surprise, then, that Mandelbrot found a better model, and in addition a simple collection of *cartoons*, basically just iterates of a broken line segment, that by varying a single parameter can be tuned to produce graphs indistinguishable from real market data. The point, of course, is not to just make *Pick the Fake* quizzes that market experts fail, though to be sure, that has some entertainment and educational value. All these cartoons have built in the self-affinity observed in real data. Pursuing the goal of constructing the most parsimonious models accounting for observation, these cartoons suggest that dependence and non-Gaussian distributions may be a consequence of properly tuned self-affinity. More detail is given in the next chapter.

Finally, these cartoons are a perfect laboratory for student experimentation.

5 Conclusion

Some view science, perhaps especially mathematics, as a serious inquiry that should remain aloof from popular culture. Many of these people regret our teaching of fractal geometry, because its images have been embraced by popular culture.

We take the opposite view. As scientists, our social responsibility includes contributing to the scientific literacy of the general population. That fractal geometry has the visual appeal to excite wide interest is undeniable. This introduction argued that fractal geometry has the substance to engage non-science students in mathematics, in a serious way and to a greater degree than any other discipline of which we are aware. The chapters of this volume amplify this position by showing how a wide variety of teachers have done this in many settings.

Chapter 2

Unsolved Problems and Still-Emerging Concepts in Fractal Geometry

Benoit B. Mandelbrot

The preceding chapter sketches a striking property of fractal geometry. Its first steps are, both literally and demonstrably, childishly easy. But high rewards are found just beyond those early steps. In particular, forbiddingly difficult research frontiers are so very close to the first steps as to be understood with only limited preparation. Evidence of this unique aspect of fractal geometry is known widely, but scattered among very diverse fields. It is good, therefore, to bring a few together. A fuller awareness of their existence is bound to influence many individuals' and institutions' perception of the methods, goals, and advancements of fractal geometry.

1 Introduction

"You find fractals easy? This is marvelous." Thus begins my response to an observation that is sometimes heard. "If you are a research mathematician, the community needs you to solve the challenging problems in this nice long list I carry around. If you are a research scientist, you could help to better analyze the important natural phenomena in this other long list."

The first half-answer is elaborated in Section 2. The point is that fractal geometry has naturally led to a number of compelling mathematical conjectures. Some took 5, 10, or 20 years to prove, others—despite the investment of enormous efforts—remain open and notorious. If anything, what slows down the growth of fractal-based mathematics is the sheer difficulty of some of its more attractive and natural portions.

The second-half answer is elaborated in Section 3. The point is that, among other features, fractal geometry is, so far, the only available language for the study of roughness, a concept that is basic and related to our senses, but has been the last to give rise to a science. In many diverse pre-scientific

fields, the absence of a suitable language delays the moment when some basic problems could be attacked scientifically. In other instances, it even delays the moment when those problems could be stated.

2 From simple visual observation to forbiddingly difficult mathematical conjecture

A resolutely purist extreme view of art holds that great achievements must be judged for themselves, irrespective of their period and the temporary failures that preceded their being perfected. In contrast, the most popular view attaches great weight to cultural context and mutual influences, and more generally tightly links the process and its end-products. Some works do not survive as being excellent but as being representative or historically important. For example, a resolutely sociological extreme view that we do not share holds that Beethoven's greatness in his time and ours distracts from the more important appreciation of his contemporaries. Few persons, and not even all teachers, are aware that a very similar conflict of views exists in mathematics.

As widely advertised, the key product of mathematics consists in theorems; in each, assumptions and conclusions are linked by a proof. It is also well known that many theorems began in the incomplete status of conjectures that include assumptions and conclusions but lack a proof. The iconic example was a conjecture in number theory due to Fermat. After a record-breaking long time it led to a theorem by Wiles. Conjectures that resist repeated attempts at a proof acquire an important role, in fact, a very peculiar one. On occasion the news that an actual proof has made a conjecture into a theorem is perceived as a letdown, while it is suggested that these conjectures' main value resides in the insights provided by both the unsuccessful and the successful searches for a proof.

Be that as it may, fractal geometry is rich in open conjectures that are easy to understand, yet represent deep mathematics. First, they did not arise from earlier mathematics, but in the course of practical investigations into diverse natural sciences, some of them old and well established, others newly revived, and a few altogether new. Second, they originate in careful inspections of actual pictures generated by computers. Third, they involve in essential fashion the century-old mathematical *monster shapes* that were for a long time guaranteed to lack any contact with the real world. Those fractal conjectures attracted very wide attention in the professions but elude proof. We feel very strongly that those fractal conjectures should not be reserved for the specialists, but should be presented to the class whenever possible. The earlier, the better. To dispel the notion that all of mathematics was done centuries ago, nothing beats being able to understand appropriate problems no one knows how to solve. Not all famous unsolved problems will work here: the Poincaré conjecture cannot be explained to high school students in an hour or a few. But many open fractal conjectures can.

For the reasons listed above, the questions raised in this chapter bear on an issue of great consequence. Does pure (or purified) mathematics exist as an autonomous discipline, one that can and should develop in total isolation from sensations and the material world? Or, to the contrary, is the existence of totally pure mathematics a myth?

The role of visual and tactile sensations. The ideal of pure mathematics is associated with the great Greek philosopher Plato (427?–347 BC). This (at best) mediocre mathematician used his great influence to free mathematics from the pernicious effects of the real world and of sensations. This position was contradicted by Archimedes (287–212 BC), whose realism I try to emulate.

Indeed, my work is unabashedly dominated by awareness of the importance of the messages of our senses. Fractal geometry is best identified in the study of the notion of roughness. More specifically, it allows a place of honor to fullfledged pictures that are as detailed as possible and go well beyond mere sketches and diagrams. Their original goal was modest: to gain acceptance for ideas and theories that were developed without pictures but were slow to be accepted because of cultural gaps between fields of science and mathematics. But those pictures then went on to help me and many others generate new ideas and theories. Many of these pictures strike everyone as being of exceptional and totally unexpected beauty. Some have the beauty of the mountains and clouds they are meant to represent; others are abstract and seem wild and unexpected at first, but after brief inspection appear totally familiar. In front of our eyes, the visual geometric intuition built on the practice of Euclid and of calculus is being retrained with the help of new technology.

Pondering these pictures proves central to a different philosophical issue. Does the beauty of these mathematical pictures relate to the beauty that a mathematician rooted in the twentieth century mainstream sees in his trade after long and strenuous practice?

2.1 Brownian clusters: fractal islands

The first example, introduced in Mandelbrot (1982), is a wrinkle on Brownian motion. The historical origins of random walk (drunkard's progress) and Brownian motion are known and easy to understand, at least qualitatively. From this, it is simple to motivate the definition of the Wiener Brownian motion: a random process $B(t)$ with increments $B(t+h) - B(t)$ that obey the Gaussian distribution of mean 0 and variance h, and that are independent over disjoint intervals.

For a given time L, the *Brownian bridge* $B_{\text{bridge}}(t)$ is defined by

$$B_{\text{bridge}}(t) = B(t) - (t/L)B(L),$$

for $0 \leq t \leq L$. Taking $B(0) = 0$, we find $B_{\text{bridge}}(L) = B_{\text{bridge}}(0)$. Combining one Brownian bridge in the x-direction and one in the y-direction and erasing time yields a *Brownian plane cluster Q*. Because we use Brownian bridges to construct it, the Brownian plane cluster is a closed curve. See Figure 1. An example of a well-known and fully proven

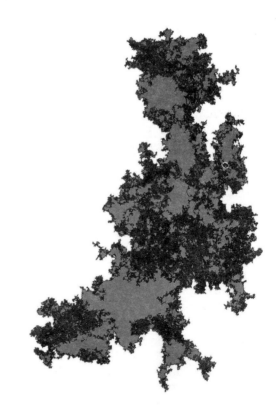

Figure 1: A Brownian plane cluster; Plate 243 of Mandelbrot (1982).

fact is that the fractal dimension of Q is $D = 2$. This result is important but not really perspicuous, because the big holes seem to contradict the association of $D = 2$ with plane-filling curves. Results that are well known and not perspicuous are not for the beginner.

Let us proceed to the *self-avoiding planar Brownian motion* $\tilde{Q}$. It is defined in Mandelbrot (1982) as the set of points of the cluster Q accessible from infinity by a path that fails to intersect Q. That is, $\tilde{Q}$ is the *hull* of Q, also called its *boundary* or *outer edge*. The hull $\tilde{Q}$ is easy to comprehend because it lacks double points. The unanswered question associated with it is the **4/3 Conjecture**, that $\tilde{Q}$ has fractal dimension 4/3.

An early example of Q, and hence of $\tilde{Q}$ is seen in Figure 1. It looks like an island with an especially wiggly coastline, and experience suggested its dimension is approximately 4/3. This comparison with islands made the 4/3 conjecture sensible and plausible in 1982 and it remains sensible and plausible to students; that it remained a conjecture for many years is something they can appreciate. Numerical tests and physicists' heuristics were added to the empirical evidence and the conjecture was proved in Lawler, Werner, & Schramm (2000).

2.2 The Mandelbrot set

Second example: In the past, music could be both popular and learned, but *elitists* believe that this is impossible today. For mathematics, the issue was not raised because no part of it could be called a part of popular culture. Providing a counterexample, no other modern mathematical object has become part of both scientific and popular culture as rapidly and thoroughly as the Mandelbrot set. Moreover, an algorithm for generating this set is readily mastered by anyone familiar with elementary algebra. Thousands of people, from middle school children to senior researchers and Fields Medalists, have written programs to visualize various aspects of the Mandelbrot set.

Recall the simplest algorithm: a complex number c belongs to the Mandelbrot set M if and only if the sequence $z_0, z_1, z_2, \ldots$ stays bounded, where $z_0 = 0$ and $z_{i+1} = z_i^2 + c$.

For instance, the sequence can stay bounded by converging to a fixed point or to a cycle. Denote by M_0 the set of all c for which this is true. Of course, $M_0 \subset M$. In fact, M_0 is of interest to the students of dynamics, hence my original investigations were of M_0, not of M. Interest shifted to M because producing pictures of M is easy. By contrast, to test if $c \in M_0$, we first generate several hundred or thousand points of the sequence $z_0, z_1, z_2, \ldots$, and test if for large enough i there is an n for which $|z_{i+n} - z_i|$ is very small. This suggests convergence to a cycle of length n. (An impractical theoretical alternative is to solve the 2^n-degree polynomial equation $f_c^n(z) = z$, where $f_c(z) = z^2 + c$, then test the stability of the n-cycle by a derivative condition:

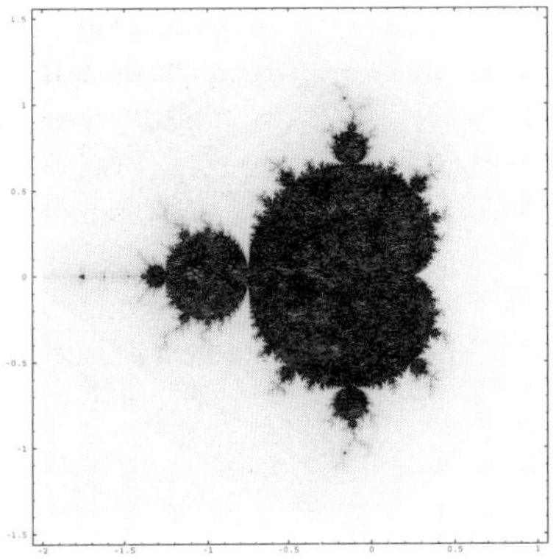

Figure 2: The Mandelbrot set.

$1 > |f_c'(w_1) \cdots f_c'(w_n)| = 2^n \cdot |w_1 \cdots w_n|$. Here the points $w_1, \ldots, w_n$ of the n-cycles are the sequences of successive z_i for different z_0. In general, for each c there are several n-cycles, but at most one is stable.)

Computer approximations of M_0 actually yield a set smaller than M_0, and computer approximations of M actually yield a set larger than M. Extending the duration of the computation seemed to make the two representations converge to each other and to an increasingly elaborate common limit. Furthermore, when c is an interior point of M, not too close to the boundary, it was easily checked that a finite limit cycle exists: the steps outlined above converge fairly rapidly for c not too close to the boundary. Those observations led me to conjecture that M is identical to M_0 together with its limits points, that is, $M = cl(M_0)$, the closure of M_0.

In terms of its being simple and understandable without any special preparation, this conjecture is difficult to top. But after almost twenty years of study, it remains a conjecture. With the proof of Fermat's last theorem, the conjecture $M = cl(M_0)$ may have been promoted to illustrating the shortest distance between a simple idea (in this case, complete with popular pictures) and deep, unsolved mathematics. (Not so simple is the usual restatement of this conjecture: that M is locally connected.)

2.3 Dimensions of self-affine sets

The first tool for quantifying self-similar fractals is dimension. For a fractal consisting of N pieces, each scaled in all directions by a factor of r, the dimension D is given by

$$D = \frac{\log(N)}{\log\left(\frac{1}{r}\right)}.$$

This is easy to motivate, trivial to compute. Working through several examples, students soon develop intuition for the visual signatures of low- and high-dimensional fractals. The generalization to self-similar fractals having different scal-

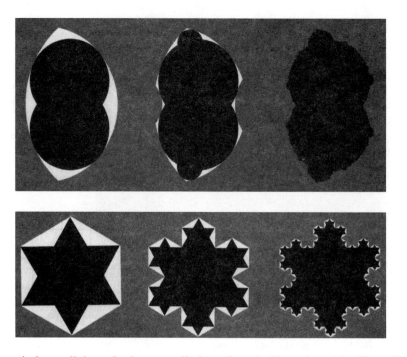

Figure 3: Top: Osculating circles outlining a Jordan curve limit set from inside and outside, Plate 177 of Mandelbrot (1982). Bottom: Osculating triangles outlining the Koch snowflake curve from inside and outisde, Plate 43 of Mandelbrot (1982).

ings for different pieces is not difficult. For a fractal consisting of N pieces, the ith piece scaled by a factor of r_i, the dimension D is the unique solution of the *Moran equation*

$$\sum_{i=1}^{N} r_i^D = 1.$$

Often this must be solved numerically, but this is not a difficulty given today's graphing calculators and computer algebra packages.

The simplicity of these calculations leads some people to believe that calculating dimensions is a simple process. This is a misperception resulting from the almost exclusive reliance on self-similar fractals for examples. The case of self-affine fractals, where the pieces are scaled by different factors in different directions, is much more difficult. Although some special cases are known, no simple variant of the Moran equation has been found. Kenneth Falconer describes the situation this way, "Obtaining a dimension formula for general self-affine sets is an intractable problem." (Falconer (1990), 129.) By simply changing the scaling factors in one direction, a completely straightforward exercise becomes tremendously difficult, perhaps without general solution.

2.4   Limit sets of Kleinian groups

A collection of Möbius transformations of the form $z \rightarrow (az + b)/(cz + d)$ defines a group that Poincaré called Kleinian. With few exceptions, their limit sets S are frac-

tal. For the closely related groups based on geometric inversions in a collection $C_1, C_2, \ldots, C_n$ of circles, there is a well-known algorithm that yields S in the limit. But it converges with excruciating slowness as seen in Plate 173 of Mandelbrot (1982). For a century, the challenge to obtain a fast algorithm remained unanswered, but it was met in many cases in Chapter 18 of Mandelbrot (1982). See also Mandelbrot (1983). In the case of this construction, fractal geometry did not open a new mathematical problem, but helped close *a very old* one.

In the new algorithm, the limit set of the group of transformations generated by inversions is specified by covering the complement of S by a denumerable collection of circles that *osculate* S. The circles' radii decrease rapidly, therefore their union outlines S very efficiently.

When S is a Jordan curve (as on Plate 177 of Mandelbrot (1982)), two collections of osculating circles outline S, respectively from the inside and the outside. They are closely reminiscent of the collection of osculating triangles that outline Koch's snowflake curve from both sides (Figure 3). Because of this analogy, the osculating construction seems, after the fact, to be very natural. But the hundred year gap before it was discovered shows it was not obvious. It came only after respectful examination of pictures of many special examples.

A particularly striking example is seen in Figure 4, called "Pharaoh's breastplate," Ken Monks' improved rendering of Plate 199 of Mandelbrot (1982). A more elaborate version of this picture appears on the cover of Mandelbrot (1999). This is the limit set of a group generated by inversion in the six

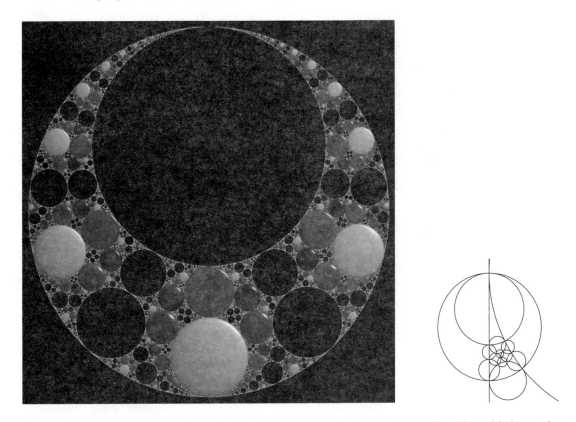

Figure 4: Left: Pharaoh's breastplate. See the color plates. Right: the six circles generating Pharaoh's breastplate, together with a few circles of the breastplate for reference.

circles drawn as thin lines on the small accompanying diagram. Here, the basic osculating circles actually belong to the limit set and do not intersect (each is the limit set of a Fuchsian subgroup based on three circles). The other osculating circles follow by all sequences of inversions in the six generators, meaning that each osculator generates a *clan* with its own *tartan* color.

By inspection, it is easy to see that this group also has three additional Fuchsian subgroups, each made of four generators and contributing full circles to the limit set.

Pictures such as Figure 4 are not only aesthetically pleasing, but they breathe new life into the study of Kleinian groups. Thurston's work on hyperbolic geometry and 3-manifolds opens up the possibility for limit sets of Kleinian group actions to play a role in the attempts to classify 3-manifolds. The Hausdorff dimension of these limit sets has been studied for some time by Bishop, Canary, Jones, Sullivan, Tukia, and others. The group G that generates the limit set gives rise to another invariant, the *Poincaré exponent*

$$\delta(G) = \inf\left\{s : \sum_{g \in G} \exp\left(-s\rho\left(0, g(0)\right)\right)\right\} < \infty$$

where ρ is the hyperbolic metric. Under fairly general conditions, the Poincaré exponent of a Kleinian group equals the Hausdorff dimension of the limit set of the group. See Bishop & Jones (1997), for example.

This is an active area of research: much remains to be done.

3 "Mathematics is a language": the emergence of new concepts

History tells us that the great Josiah Willard Gibbs (1839–1903) made this remark at a Yale College Faculty meeting devoted to the reform of foreign language requirements (some faculty issues never die!). The context may seem undignified or amusing, but, in fact, Gibbs's words bring forth a deep issue. To express subtle scientific ideas, one often needs new *words* that are subtler than those of ordinary language.

As background, everyone knows that some great books deservedly became classics because they provided, for the first time, a new language in which personal emotions—that the reader would feel but not be able to express—could be both refined and made public. This is not at all a matter of coining new words for old concepts but of making altogether new concepts emerge.

Advances in the sciences are assessed in diverse ways, one of which is the emergence of new scientific concepts. Indeed, the facile precept that the first step is to observe then measure, sounds less compelling when the object of study is an undescribable mess and all the measurements that readily come to mind disagree or even seem self-contradictory. This is why the point of passage from prescientific to scientific investigation is often marked by what Thomas Kuhn called *change of paradigm*. Sometimes this includes the appearance of a suitable new language, without which observations could not be made and quantified.

3.1 Fractals are a suitable language for the study of roughness wherever it is encountered

Let us ponder the ubiquity of the notion of roughness and its lateness in becoming formalized. Many sciences arose directly from the desire to describe and understand some basic messages the brain receives from the senses. Visual signals led to the notions of bulk and shape and of brightness and color. The sense of heavy versus light led to mechanics and the sense of hot versus cold led to the theory of heat. Other signals (for example, auditory) require no comment. Proper measures of mass and size go back to prehistory and temperature, a proper measure of hotness, dates to Galileo.

Against this background, the sense of smooth versus rough suffered from a level of neglect that is noteworthy though hardly ever pointed out. Not only does the theory of heat have no parallel in a theory of roughness, but temperature itself had no parallel until the advent of fractal geometry. For example, in the context of metal fractures, roughness was widely measured by a root mean square deviation from an interpolating plane. In other words, metallurgists used the same tool as finance experts used to measure volatility. But this measurement is inconsistent. Indeed, different regions of a presumably homogeneous fracture emerged as being of different r.m.s. volatility. The same was the case for different samples that were carefully prepared and later broken following precisely identical protocols.

To the contrary, as shown in Mandelbrot, Passoja & Paullay (1984) and confirmed by every later study, the fractal dimension D, a characteristic of fractals, provides the desired invariant measure of roughness. The quantity $3 - D$ is called the codimension or Hölder exponent by mathematicians and now called the roughness exponent by metallurgists.

The role played by exponents must be sketched here. It is best in this chapter to study surfaces through their intersections by approximating orthogonal planes. Had these functions been differentiable, they could be studied through the derivative defined by $P'(t) = \lim_{\epsilon \to 0}(1/\epsilon)[P(t + \epsilon) - P(t)]$. For fractal functions, however, this limit does not exist and the local behavior is, instead, studied through the parameters of a relation of the form $dP \sim F(t)(dt)^\alpha$. Here $F(t)$ is called the prefactor, but the most important parameter is the exponent $\alpha = \lim_{\epsilon \to 0}\{\log[P(t + \epsilon) - P(t)]/\log \epsilon\}$.

There is an adage that, when you own only a hammer, everything begins to look like a nail. This adage does not apply to roughness.

3.2 Fractals and multifractals in finance

Versions of the Brownian motion model mentioned in Section 2.1 are widely used to model aspects of financial markets. In fact, and contrary to common belief, the first analysis of Brownian motion was not advanced in 1905 by Einstein. In 1900 Bachelier had already developed Brownian motion to study the stock market.

Despite this historical precedent, successive differences of real data sampled at equal time intervals reveal even on cursory investigation that Brownian models are very far from being tolerable. Most visibly, (1) the width of the *central band* is not constant, but varies substantially, (2) the excursions from the central band are so large as to be astronomically unlikely in the Brownian case, and (3) the excursions are not independent, but occur in clumps, often when the underlying band is widest. Figure 5 illustrates these differences.

Ad hoc fixes can account for each of these failures of the Brownian model, but very rapidly become far too complicated for anybody, especially for courses not addressed to experts. The fractal/multifractal approach of Mandelbrot (1997) is much more elegant. It provides a unified way to synthesize all, and moreover introduces a family of parameterized cartoon models suitable for student exploration.

Let us dwell on what is happening. Compared with well-developed standard mathematical finance, the fractal cartoons are incomparably more satisfactory. But they are far simpler than the first stages of standard finance, so simple that they have been immediately incorporated into both *Fractal Geometry for Non-Science Students* (a course primarily for humanities students) and *Fractal Geometry: Techniques and Applications* (a course for sophomore-junior math and science students) at Yale. In effect, students are invited to participate in discussions between experts. They are amazed by the realistic appearance of forgeries made with these cartoons. Showing the class a collection of real data and forgeries always produces interesting results. Students disagree, sometimes with great animation, about which are real and which are forgeries. The *inverse problem*, finding a cartoon to create a forgery of a particular data set, has been a source of interesting student projects, some quite creative. After studying background in the different visual signatures of long tails and global dependence, students are amazed at how slight changes in the cartoon generator can achieve both effects.

The basic construction of the cartoon involves an *initiator* and a *generator*. The process to be iterated consists of replacing each copy of the initiator with an appropriately rescaled copy of the generator. For a first cartoon, the initiator is the diagonal of the unit square, and the generator is the broken line with vertices $(0, 0)$, $(4/9, 2/3)$, $(5/9, 1/3)$, and $(1, 1)$. Fig-

Figure 5: Left: differences in successive daily closing prices for four years of EMC data. Right: successive differences of the same number of steps in one-dimensional Brownian motion.

Figure 6: The initiator, generator, and first iterate of a non-random Brownian motion cartoon.

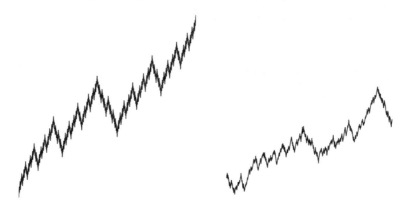

Figure 7: Left: the 6th iterate of a non-random Brownian cartoon. Right: the 6th iterate of a randomized Brownian cartoon.

ure 6 shows the initiator (left), generator (middle), and first iteration of the process (right).

To get an appreciation for how quickly the jaggedness of these cartoons grows, the left side of Figure 7 shows the 6th iterate of the process.

Self-affinity is guaranteed because it is built into the process; each piece is an appropriately scaled version of the whole. In this case, the scaling ratios have been selected to satisfy the square root condition of Brownian motion. The horizontal axis denotes time t, the vertical denotes price x. The first and third generator segments have $\Delta t_1 = \Delta t_3 = \frac{4}{9}$ and $\Delta x_1 = \Delta x_3 = \frac{2}{3}$; the middle segment has $\Delta t_2 = \frac{1}{9}$ and $\Delta x_2 = -\frac{1}{3}$. So for each generator segment we have $|\Delta x_i| = (\Delta t_i)^{1/2}$.

A cartoon is *unifractal* if there is a constant H so that for each generator segment $|\Delta x_i| = (\Delta t_i)^H$. If different H are needed for different segments, the cartoon is *multifractal*.

The left side of Figure 7 is far too regular to mimic any real data. But it can be randomized easily by shuffling the order in which the three pieces of the generator are put into each

scaled copy. The right side of Figure 7 shows the result of this shuffling, for the sixth stage of the construction.

Instead of the graph itself, it is less common but far better to look at the increments. The cartoon sequence we have produced has jumps at uneven intervals: some at multiples of $1/3^n$, some at multiples of $1/9^n$. Because we rarely have detailed knowledge of the underlying dynamics generating real data, measurements usually are taken at equal time steps. To construct a sequence of appropriate increments, we sample the graph at fixed time intervals and subtract successive values obtained. Operationally, first make a list of time values for the sampling, then find the cartoon time values between which each sample value lies, and linearly interpolate between the cartoon values to find the sample value at the sample time.

Figure 8 illustrates how the statistical properties of the differences can be modified by making a simple adjustment in the generator. Fixing the points $(0, 0)$ and $(1, 1)$, we keep the middle turning points symmetrical: $(a, \frac{2}{3})$ and $(1 - a, \frac{1}{3})$, where a lies in the range $0 < a \le \frac{1}{2}$. All pictures were constructed from the tenth generation, hence consist of $3^{10} =$

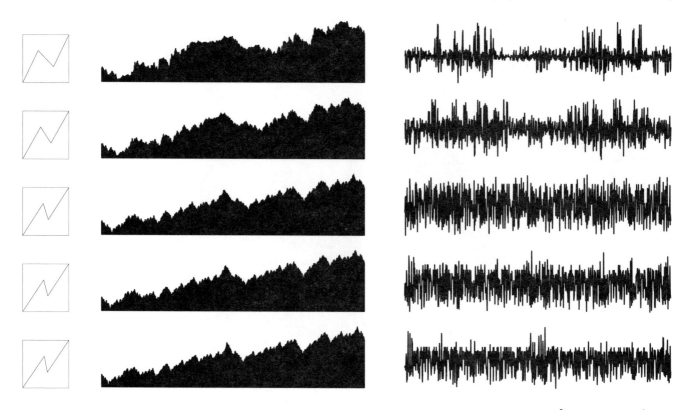

Figure 8: Generators, cartoons and difference graphs for symmetric cartoons with turning points $(a, \frac{2}{3})$ and $(1 - a, \frac{1}{3})$, for $a = 0.333, 0.389, 0.444, 0.456$, and 0.467. The same random number seed is used in all graphs.

59,049 intervals. The difference graphs are constructed by sampling at 1000 equal time steps.

Certainly, correlations are introduced as the point $(4/9, 2/3)$ is moved to the left. Of course, this is just the beginning. More detailed study reveals relations between the Hölder exponents and the slopes of the generator intervals, and properties of the multifractal measure can be extracted from the cartoons (Mandelbrot (1997)). The H-exponents and the $f(\alpha)$ curve are much too technical for *Fractal Geometry for Non-Science Students*, but are appropriate topics for the more mathematically sophisticated *Fractal Geometry: Techniques and Applications*. Even for this audience, these are challenging concepts. Yet these simple cartoons provide accessible introductions to some of the subtle mathematics of multifractals.

As a last example, we mention a fascinating theorem and a visual representation of its meaning. The Yale students taking *Fractal Geometry for Non-Science Students* in autumn of 1998 followed the development of Figure 9 with passion and helped improve it. The generator increments Δt represent *clock time*. Viewed in clock time, prices sometimes remain quiescent for long periods, and sometimes change with startling rapidity, perhaps even discontinuously. For these cartoons, clock time can be recalibrated to uniformize these changes in price variation. Basically, slow the clock during periods of rapid activity and speed it during periods of low activity. Students found the VCR a useful analog. Fast-forward through the commercials (low activity) and use slow-motion through the interesting bits (rapid activity).

For the cartoon generators, this is achieved by first finding the unique solution D of $|\Delta x_1|^D + \cdots + |\Delta x_n|^D = 1$, then defining the *trading time* generators by $\Delta T_i = |\Delta x_i|^D$. By changing to trading time, every multifractal price cartoon can be converted into a unifractal cartoon in multifractal time. Global dependence and long tails are unpacked in different ways by converting to trading time. Specifically, global dependence remains in the price vs. trading time record, but the long tails are absorbed into the multifractal nature of trading time.

Figure 9 shows a three-dimensional representation of this conversion. Note how the clock time-trading time curve compresses the flat regions and expands the steep regions of the price-clock time graph. Thus the long tails of the price-clock time graph are absorbed into the multifractal time measure. In addition, the dependence of increments is uniformized to fractional Brownian motion in the price-trading time graph. That is, the conversion to trading time decomposes long tails and dependent increments into different aspects of the graph.

Starting from a rough idea of such a representation, this picture evolved over about a week, through discussions with the class. Few things have excited the class as much as being

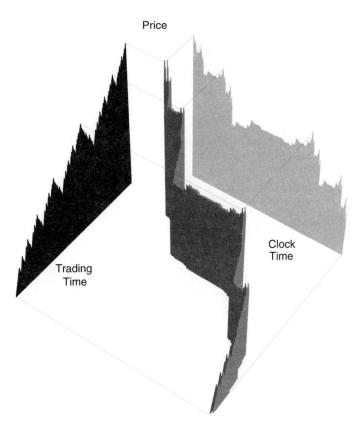

Figure 9: Converting the price-clock time graph to the price-trading time graph by means of the clock time-trading time graph.

involved, as a group, in the production of a figure to explain current research in the field.

4   Conclusion

A famous tongue-twister and test in Greek and evolution, due to E. H. Haeckel, proclaims that "ontogeny recapitulates phy-logeny." In plainer English, the early growth of an individual repeats the evolution of its (his, her) ancestors. As argued elsewhere in this book (Chapter 3), this used to be the **BIG PICTURE** historical justification of *old math*—not a well-documented one. But teachers ought to welcome any well-documented **small picture** version that happens to come their way. Fractals deserve to be welcomed.

Chapter 3

Fractals, Graphics, and Mathematics Education[1]

Benoit B. Mandelbrot

1 Introduction

The fundamental importance of education has always been very clear to me and it has been very frustrating, and certainly not a good thing in itself, that the bulk of my working life went without the pleasures and the agonies of teaching. On the other hand, there is every evidence, in my case, that being sheltered from academic life has often been a necessary condition for the success of my research. An incidental consequence is that some of the external circumstances that dominated my life may matter to the story to be told here, and it will be good to mention them, in due time.

But past frustrations are the last thing to dwell upon in this book. Watching some ideas of mine straddle the chasm between the research frontier and the schools overwhelms me with a feeling of deep accomplishment. Clearly, for better or worse, I have ceased to be alone in an observation, a belief, and a hope, that keep being reinforced over the years.

The observation is that fractals—together with chaos, easy graphics, and the computer—enchant many young people and make them excited about learning mathematics and physics. In part, this is because an element of instant gratification happens to be strongly present in this piece of mathematics called fractal geometry. The belief is that this excitement can help make these subjects easier to teach to teenagers and to beginning college students. This is true even of those students who do not feel they will need mathematics and physics in their professions. This belief leads to a hope—perhaps megalomaniac—concerning the abyss which has lately separated the scientific and liberal cultures. It is a cliché, but

one confirmed by my experience, that scientists tend to know more of music, art, history, and literature, than humanists know of any science. A related fact is that far more scientists take courses in the humanities than the other way around. So let me give voice to a strongly held feeling. An element of instant gratification happens to be strongly present in this piece of mathematics called fractal geometry. Would it be extravagant to hope that it could help broaden the small band of those who see mathematics as essential to every educated citizen, and therefore as having its place among the liberal arts?

The lost unity of liberal knowledge is not just something that old folks gather to complain about; it has very real social consequences. The fact that science is understood by few people other than the scientists themselves has created a terrible situation. One aspect is a tension between conflict of interest and stark ignorance: that vital decisions about science and technology policy are all too often taken either by people so closely concerned that they have strong vested interests, or by people who went through the schools with no math or science. Thus, every country would be far better off if understanding and appreciation for some significant aspect of science could become more widespread among its citizens. This demands a liberal education that includes substantial instruction in math.

Fractals prove to have many uses in technical areas of mathematics and science. However, this will not matter in this chapter. Besides, if fractals' usefulness in teaching is confirmed and proves lasting, this is likely to dwarf all their other uses.

This chapter shall assume all of you to have a rudimentary awareness or knowledge of fractals, or will one day become motivated to acquire this knowledge elsewhere. My offering is my book, Mandelbrot (1982), but there are many more

---

[1] Adapted from a closing invited address delivered at the Seventh International Congress of Mathematics Education (ICME-7), held in 1992 at Laval University of Quebec City. The text remains self-contained and preserves some of the original flavor; it repeats some points that were already used elsewhere in this book but bear emphasis.

sources at this point. For example, the website

$$\text{http://classes.yale.edu/math190a/}$$
$$\text{Fractals/Welcome.html}$$

is a self-contained short course on basic fractal geometry.

I shall take up diverse aspects of a basic and very concrete question about mathematics education: what should be the relations—if any—between (a) the overall development of mathematics in history, (b) the present status of the best and brightest in mathematics research, and (c) the most effective ways of teaching the basics of the field?

2   Three mutually antagonistic approaches to education

By simplifying (strongly but not destructively), one can distinguish three mutually antagonistic approaches to mathematical education. The first two are built on *a priori* doctrine: the **old math**, dominated by (a) above, and the **new math**, dominated by (b). (I shall also mention a transitional approach between old and new math.) To the contrary, the approach I welcome would be resolutely pragmatic. It would encourage educational philosophy to seek points of easiest entry. In this quest, the questions of how mathematics research began and of its present state, are totally irrelevant.

To elaborate by a simile loaded in my favor, think of the task of luring convinced nomads into hard shelter. One could tempt them into the kinds of shelters that have been built long ago, in countries that happened to provide a convenient starting point in the form of caves. One could also try to tempt them into the best possible shelters, those being built far away, in highly advanced countries where architecture is dominated by structurally pure skyscrapers. But both strategies would be most ill-inspired. It is clearly far better to tempt our nomads by something that interests them spontaneously. But such happens precisely to be the case with fractals, chaos, easy graphics, and the computer. Hence, if their effectiveness becomes confirmed, a working pragmatic approach to mathematics education may actually be at hand. We may no longer be limited to the old and new math. Let me dwell on them for a moment.

2.1   The old math approach to mathematics education

The old math approach to mathematics education saw the teacher's task as that of following history. The goal was to guide the child or young person of today along a simplified sequence of landmarks in the progress of science throughout the history of humanity. An extreme form of this approach prevailed until mid-nineteenth century in Great Britain, the

sole acceptable textbook of geometry being a translation of Euclid's *Elements*.

The folk-psychology behind this approach asserted with a straight face that the mental evolution of mankind was the product of historical necessity and that the evolution of an individual must follow the same sequence. In particular, the acquisition of concepts by the small child must follow the same sequence as the acquisition of concepts by humankind. Piaget taught me that such is indeed the case for concepts that children must have acquired before they start studying mathematics.

All this sounds like a version of "ontogeny recapitulates phylogeny," but it is safe to say that people had started developing mathematics well before Euclid. As a matter of fact, those who edited the *Elements* were somewhat casual and left a number of propositions in the form of an archaeological site where the latest strata do not completely hide some tantalizing early ones. To be brief, what we know of the origin of mathematics is too thin and uncertain to help the teacher.

Be that as it may, an acknowledged failing of old math was that the teacher could not conceivably move fast enough to reach modern topics. For example, the school mathematics and science taken up between ages 10 and 20 used to be largely restricted to topics humanity discovered in antiquity. As might be expected, teachers of old lit heard the same criticism. A curriculum once reserved to Masters of Antiquity was gradually changed to leave room for the likes of Shakespeare and of increasingly modern authors; in the USA, it had to yield room to American Masters, then to multicultural programs.

2.2   Transitional approaches to mathematics education

Concerns about old math are an old story. Consider two examples. In Great Britain, unhappiness with Euclid's *Elements* as a textbook fueled the reforms movement that led in 1871 to the foundation of the *Association for the Improvement of Geometrical Teaching* (in 1897 it was renamed the *Mathematical Association*). As a student in France around 1940, I heard about a reform movement that had flourished before 1900. It motivated Jacques Hadamard (1865–1963), a truly great man, to help high school instruction by writing Hadamard (1898), a modern textbook of geometry that stressed the notion of transformation. I was given a copy and greatly enjoyed it, but the consensus was that it was far above the heads of those it hoped to please. In Germany, there was the book by Hilbert and Cohn–Vossen (1952).

But the 20th century witnessed a gradual collapse of geometry. Favored topics became arithmetic and number theory; they have ancient roots, are one of the top fields in today's mathematical research, and include large portions that are independent of the messy rest of mathematics. Therefore, they are central to many charismatic teachers' efforts to fire youngsters' imaginations towards mathematics.

2.3 The 1960s and the new math approach to mathematics education

Far bolder than those half-hearted attempts to enrich the highly endowed students with properly modern topics was the second broad approach to mathematics education exemplified by the new math of the 1960s.

Militantly anti-historical, I viewed the state of mathematics in the 1960s, and the direction in which it was evolving at that particular juncture in history, as an intrinsic product of historical necessity. This is what made it a model at every level of mathematics education. If the research frontier of the 1960s had *not* been historically necessary, new math would have lost much of its gloss or even legitimacy. The evidence, however, is that the notion of historical necessity as applied to mathematics (as well as other areas!) is merely an ideological invention. This issue is important and tackled at length in Chapter 4 of this volume.

In any event, new math died a while ago, victim of its obvious failure as an educational theory. The Romans used to say that "of the dead, one should speak nothing but good." But the new math's unmitigated disaster ought at least teach us how to avoid a repetition. However, it is well known that failure is an orphan (while success has many would-be parents), that is, no responsibility for this historical episode is claimed by anyone, as of today.

Take for example the French formalists who once flourished under the pen-name of Bourbaki (I shall have much more to say about them). They nurtured an environment in which new math became all but inevitable, yet today they join everyone else in making fun of the outcome, especially when it hurts their own children or grandchildren. This denial of responsibility is strikingly explicit in a one-hour story a French radio network devoted to the Bourbaki a few years ago. (Audio-cassettes may be available from the Société Mathématique de France.) One hears in it that the Bourbaki bear no more responsibility than the French man in the street (failure is indeed an orphan), and that they have never made a statement in favor of new math. On the other hand, having paid attention while suffering through the episode as the father of two sons, I do not recall their making a statement against new math, and I certainly recall the mood of that time.

Be that as it may, it is not useful to wax indignant, but important to draw a lesson for the future. The lesson is that *no frontier mathematics research must again be allowed to dominate mathematics education.* At the other unacceptable extreme, needless to say, I see even less merit in the notion that one can become expert at teaching mathematics or at writing textbooks, yet know nothing at all about the subject. Quite to the contrary, the teachers and the writers must know a great deal about at least some aspects of mathematics.

Fortunately, mathematics is not the conservatives' ivory tower. As will be seen in Chapter 4, I see it as a very big house that offers teachers a rich choice of topics to study and transmit to students. The serious problem is how to choose among those topics. My point is that this choice must not be left to people who have never entered the big house of mathematics, nor to the leaders of frontier mathematics research, nor to those who claim authority to interpret the leaders' preferences. Of course, you all know already which wing of the big house I think deserves special consideration. But let me not rush to talk of fractals, and stop to ask why the big house deserves to be visited.

3   The purely utilitarian argument for widespread literacy in mathematics and science

My own experiences suggest, and all anecdotal reports confirm, that traditional mathematics (of the kind described in the section before last) does marvels when a very charismatic teacher meets ambitious and mathematically gifted children. Helping the very gifted and ambitious is an extraordinarily important task, both for the sake of those individuals and of the future development of math and science. But (as already stated) I also believe that math and science literacy must extend beyond the very gifted pupils.

Unfortunately, as we all know, this belief is not shared by everyone. How can we help it become more widely accepted? All too often, I see the need for math and science literacy referred to exclusively in terms of the needs (already mentioned) of future math and science teaching and research, and those of an increasingly technological society. To my mind, however, this direct utilitarian argument fails on two accounts: it is not politically effective; and it is not sufficiently ambitious.

First of all, if scientific literacy is valuable and remains scarce, it has always been hard to explain why the scientifically literate fail (overall) to reap the financial rewards of valuable scarcity. In fact, scientific migrant workers, like agricultural ones, keep pouring in from poorer countries. Recent years were especially unkind to the utilitarian argument since many engineers and scientists are becoming unemployed and had to move on to fields that do not require their specialized training.

Even though this is an international issue, allow me to center the following comments on the conditions in the USA. In its crudest form, very widespread only a few years ago, the utilitarian argument led many people to compare the United States unfavorably to countries, including Russia, France, or Japan, with far more students in math or science. Similarly unfavorable comparisons concerned foreign language instruction in the USA to that in other countries. The explanation in the case of the languages of Hungary or Holland is obvious: the Hungarians are not genetically or socially superior to the Austrians, but the Austrians speak German, a useful language, while Hungarian is of no use elsewhere; hence, multilingual Hungarians receive unquestioned real-life rewards. Similarly, school programs heavy in compulsory math are tolerated in France and Japan because they provide unquestioned great real-life rewards to those who do well in math.

For example, many jobs in France that require little academic knowledge to be performed are reserved (by law) for those who pass a qualifying examination. The exam seeks objectivity, and ends up being heavy on math. There are many applicants, the exams are difficult, and the students are motivated to be serious about preparing for them.

Some of these jobs are among the best possible. For example, in many French businesses one cannot approach the top unless one started at the Ecole Polytechnique, the school I attended. (I first entered Ecole Normale Supérieure, but left immediately). For a time after Polytechnique was founded (in 1794), it first selected and judged its students on the broad and subjective grounds ideally used in today's America, but later the criteria for entrance and ranking became increasingly objective—that is, mathematical. One reason was the justified fear of nepotism and political pressure, another the skill of Augustin Cauchy (1789–1857), a very great mathematician and also a master at exerting self-serving political pressure.

The result was clear at the forty-fifth and fiftieth reunions of my class at Polytechnique. For a few freshly retired classmates a knowledge of science had been essential. But most had held very powerful positions to general acclaim, yet hardly remembered what a complex number is—because it has not much mattered to them. They gave no evidence of an exceptionally strong love of science. (I do not know what to make of the number of articles our *Alumni Monthly* devotes to the paranormal.) But my classmates could never have reached those powerful positions without joining the Polytechnique "club"; to be a wizard at math, at least up to age twenty, was part of the initiation and a desired source of homogenity.

The United States of America also singles out an activity that brings monetary rewards and prestige that continue through a person's life—independent of the person's profession. This activity is sports. In France it is math. For example, one of my classmates (Valéry Giscard d'Estaing) became President of France, his goal since childhood; to help himself along, he chose to go to a college even more demanding than MIT.

For a long time France recognized a second path to the top: a mastery of Greek or Latin writers and philosophers. But by now this path has been replaced by an obstacle course in public administration. A competition continues between the two ways of training for the top, but no one claims that either mathematics or the obstacle courses is important *per se*. You see how little bearing this French model has on the situation in the USA.

Needless to say, many French people have always complained that their school system demands more math than is sensible; other French people complain that the teaching of math is poor. And I heard the same complaints on a trip to Japan. So my feeling is that the real problem may not involve embarrassing national comparisons.

4   In praise of widespread literacy in mathematics and science

Lacking the purely utilitarian argument, what could one conceivably propose to justify more and better math and physics? When I was young some of my friends were delighted to reserve real math to a small elite. But other friends and I envied the historians, the painters, and the musicians. Their fields also involved elite training, yet their goals seemed blessed by the additional virtue of striking raw nerves in other human beings. They were well understood and appreciated by a wide number of people with comparatively minimal and unprofessional artistic education. To the contrary, the goals of my community of mathematicians were becoming increasingly opaque beyond a circle of specialists. Tongue in cheek, my youthful friends and I dreamt of some extraordinary change of heart that would induce ordinary people to come closer to us of their own free will. They should not have to be bribed by promises of jobs and money, as was the case for the French adolescents. Who can tell, a popular wish to come closer to us might even induce them to buy tickets to our performances!

When our demanding dream was challenged as ridiculous and contrary to history and common sense, we could only produce one historical period when something like our hopes had been realized. Our example is best described in the following words of Sir Isaiah Berlin (Berlin 1979):

"Galileo's method ... and his naturalism, played a crucial role in the development of seventeenth-century thought, and extended far beyond technical philosophy. The impact of Newton's ideas was immense: whether they were correctly understood or not, the entire program of the Enlightenment, especially in France, was consciously founded on Newton's principles and methods, and derived its confidence and its vast influence from his spectacular achievements. And this, in due course, transformed—indeed, largely created—some of the central concepts and directions of modern culture in the West, moral, political, technological, historical, social— no sphere of thought or life escaped the consequences of this cultural mutation. This is true to a lesser extent of Darwin.... Modern theoretical physics cannot, has not, even in its most general outlines, thus far been successfully rendered in popular language as Newton's central doctrines were, for example, by Voltaire."

Voltaire was, of course, the most celebrated French writer of the eighteenth century and mention of his name brings to mind a fact that is instructive but little known, especially outside France: it concerns the first translation of Newton into French, which appeared in Voltaire's time. Feminists, listen: the translator was Gabrielle Émilie le Tonnelier de Breteuil, marquise du Châtelet-Lomont (1706–49). Madame du Châtelet was a pillar of High Society: her salon was among the most brilliant in Paris.

In addition, the XVIIIth century left us the letters that the great Leonhard Euler (1707–83) wrote to "a German princess" on topics of mathematics. Thus a significantly broad scientific literacy was welcomed and conspicuously present in a century when it hardly seemed to matter.

5 Contrasts between two patterns for hard scientific knowledge: Astronomy and history

Why is there such an outrageous difference between activities that appeal to many (like serious history), and those which only appeal to specialists? To try and explain this contrast, let me sketch yet another bit of history, comparing knowledge patterned after astronomy and history.

The Ancient Greeks and the medieval scholastics saw a perfect contrast between two extremes: the purity and perfection of the Heaven, and the hopeless imperfection of the Earth. *Pure* meant subject to rational laws which involve simple rules yet allow excellent predictions of the motion of planets and stars. Many civilizations and individuals believe that their lives are written up in full detail in a book and hence can in theory be predicted and cannot be changed. But many others (including Ancient Greeks) thought otherwise. They expected almost everything on Earth to be a thorough mess. This allowed events that were in themselves insignificant to have unpredictable and overwhelming consequences—a rationalization for magic and spells. This *sensitive dependence* became a favorite theme of many writers; Benjamin Franklin's *Poor Richard's Almanac* (published in 1757), retells an ancient ditty as follows:

> "A little neglect may breed mischief.
> For lack of a nail, the shoe was lost;
> for lack of a shoe, the horse was lost;
> for lack of a horse, the rider was lost;
> for lack of a rider, the message was lost;
> for lack of a message, the battle was lost;
> for lack of a battle, the war was lost;
> for lack of a war, the kingdom was lost;
> and all because of one horseshoe nail."

From this perspective, it seems to me that belief in astrology, and the hopes that continue to be invested today in diverse would-be sciences, all express a natural desire to escape the terrestrial confusion of human events and emotions by putting them into correspondence with the pure predictability of the stars.

The beautiful separation between pure and impure (confused) lasted until Galileo. He destroyed it by creating a terrestrial mechanics that obeyed the same laws as celestial mechanics; he also discovered that the surface of the Sun is covered with spots and hence is imperfect. His extension of the domain of order opened the route to Newton and to science. His extension of the domain of disorder made our vision of the universe more realistic, but for a long time it removed the Sun's surface from the reach of quantitative science.

After Galileo, knowledge was free from the Greeks' distinction between Heaven and Earth, but it continued to distinguish between several levels of knowledge. At one end was *hard* knowledge, a science of order patterned after astronomy. At the other end, is *soft* knowledge patterned after history, i.e., the study of human and social behavior. (In German, the word *Wissenschaft* stands for both *knowledge* and *science*; this may be one of several bad reasons why the English and the French often use *science* as a substitute for *knowledge*.)

Let me at this point confess to you the envy I experienced as a young man, when watching the hold on minds that is the privilege of psychology and sociology, and of my youthful dreams of seeing some part of hard science somehow succeed in achieving a similar hold. Until a few decades ago, the nature of science made this an idle dream. Human beings (not all, to be sure, but enough of them) view history, psychology, and sociology as alive (unless they had been smothered by mathematical modeling). Astronomy is not viewed as alive; the Sun and the Moon are superhuman because of their regularity, therefore gods. In the same spirit, many students view math as cold and dry, something wholly separate from any spontaneous concern, not worth thinking about unless they are compelled. Scientists and engineers must know the rules that govern the motions of planets. But these rules have limited appeal to ordinary humans because they have nothing to do with history or the messy, everyday life, in which, let me repeat, the lack of a nail can lose a horse (a battle, a war, and even a kingdom) or a bride.

6 A new kind of science: Chaos and fractals

Now we are ready for my main point. In recent years the sharp contrast between astronomy and history has collapsed. We witness the coming together, not of a new *species* of science; nor even (to continue in taxonomic terms) a new *genus* or *family*, but a much more profound change. Towards the end of the 19th century, a seed was sowed by Poincaré and Hadamard; but practically no one paid attention, and the seed failed to develop until recently. It is only since the 1960s that the study of true disorder and complexity has come onto

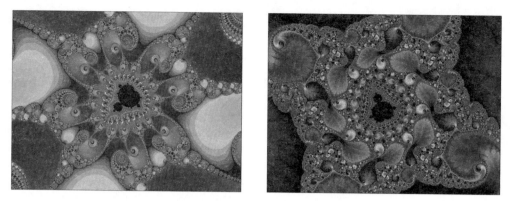

Figure 1: Two close-up views of parts of the Mandelbrot set.

the scene. Two key words are *chaos* and *fractals*, but I shall keep to *fractals*. Again and again my work has revealed cases where simplicity breeds a complication that seems incredibly lifelike.

The crux of the matter is a geometric object that I first saw in 1979, took very seriously, and worked hard to describe in 1980. It has been named the *Mandelbrot set*. It starts with a formula so simple that no one could possibly have expected so much from it. You program this silly little formula into your trusty personal computer or workstation, and suddenly everything breaks loose. Astronomy described simple rules and simple effects, while history described complicated rules and complicated effects. Fractal geometry has revealed simple rules and complicated effects. The complication one sees is not only most extraordinary but is also spontaneously attractive, and often breathtakingly beautiful. See Figure 1. Besides, you may change the formula by what seems a tiny amount, and the complication is replaced by something altogether different, but equally beautiful.

The effect is absolutely like an uncanny form of white magic. I shall never forget the first time I experienced it. I ran the program over and over again and just could not let it go. I was a visiting professor at Harvard at the time and interest in my pictures immediately proved contagious. As the bug spread, I began to be stopped in the halls by people who wanted to hear the latest news. In due time, the *Scientific American* of April 1985 published a story that spread the news beyond Harvard.

The bug spread to tens, hundreds, and thousands of people. I started getting calls from people who said they loved those pictures so much that they simply had to understand them; where could they find out about the multiplication of complex numbers? Other people wrote to tell me that they found my pictures frightening. Soon the bug spread from adults to children, and then (how often does this happen?) from the kids to teachers and to parents.

Lovable! Frightening! One expects these words to be applied to live, warm bodies, not to mere geometric shapes. Would you have expected kids to go to you, their teachers,

and ask you to explain a mathematical picture? And be eager enough to volunteer to learn more and better algebra? Would you expect strangers to stop me in a store downtown, because they just have to find out what a complex number is?

Next, let me remind you that the new math fiasco started when a committee of my elders, including some of my friends, all very distinguished and full of goodwill, figured out among themselves that it was best to start by teaching small kids the notions that famous professors living in the 1950s viewed as being fundamental, and therefore simple. They wanted grade schoolers to be taught the abstract idea of a set. For example, a box containing five nails was given a new name: it became a set of five nails. As it happened, hardly anyone was dying to know about five-nail sets.

On the other hand, the initial spread of fractals among students and ordinary people was neither planned nor supported by any committee or corporation, least of all by IBM, which supported my scientific work but had no interest in its graphic or popular aspects. This spread was one of the most truly spontaneous events I ever heard of or witnessed. People could not wait to understand and master the white magic and find out about those crazy Mandelbrot sets. The five-nail set was rejected as cold and dry. The Mandelbrot set was welcomed almost as if it were alive. Everything suggests that its study can become a part of liberal knowledge!

Chaotic dynamics meets the same response. There is no fun in watching a classical pendulum beat away relentlessly, but the motion of a pendulum made of two hinged sticks is endlessly fascinating. I believe that this contrast reveals a basic truth that every scientist knows or suspects, but few would concede. The only trace of historical necessity in the evolution of science may be that its grand strategy is to begin with questions that are not necessarily the most exciting, but are simple enough to be tackled at a given time.

The lesson for the educator is obvious. Motivate the students by that which is fascinating, and hope that the resulting enthusiasm will create sufficient momentum to move them through material that must be studied but is less widely viewed as fun.

7 Just beyond the easy fractals lurk overwhelming challenges

This last word, "fun," deserves amplification. The widely perceived difficulty of mathematics is a reason for criticism by the outsider. But for the insider it is a source of pride, and mathematics is not viewed as real unless it is difficult. In that sense, fractal geometry is as real as can be, but with a few uncommon wrinkles.

The first uncommon wrinkle has already been mentioned: hardly any other chapter of mathematics can boast that even to the outsider its first steps are fun.

Pushing beyond the first steps, a few additional ones led me (and soon led others beyond counting) to stunning observations that the eye tells us must be true, but the mind tells us must be proven.

A second uncommon wrinkle of fractal geometry is that those observations are often both simple and new; at least, they are very new within recent memory. Hardly any other chapter of mathematics can boast of simple and new observations worth making. Therefore, fractal geometry has provided multitudes with the awareness that the field of mathematics is alive.

A third uncommon wrinkle of fractal geometry is that, next to simple and new observations that were easy to prove, several revealed themselves beyond the power of the exceptionally skilled mathematicians who tackled them. Thus, some of my earliest observations about the Mandelbrot set remain open. Furthermore, no one knows the dimensions of self-affine sets beyond the simplest. In physics, turbulence and fractal aggregates remain mysterious. The thrills of frontier life can be enjoyed right next to the thriving settlements. Hardly any other chapter of mathematics can boast of so many simple but intractable conjectures.

A fourth wrinkle concerns the easy beginnings of fractal geometry. Thanks to intense exposure, it is quite true that much about fractals appears obvious today. But yesterday the opposite view was held by everyone. My writings have—perhaps with excessive verve—blamed mathematicians for having boxed themselves and everyone else in an intellectual environment where constructions now viewed as *proto-fractal* were once viewed as *pathological* and anything but obvious. This intellectual environment was proud of having broken the connections between mathematics and physics. Today there is a growing consensus that the continuity of the links between mathematics and physics is obvious, but the statements ring false in the mouths of those who denied and destroyed this continuity; they sound better in the mouths of those who rebuilt it.

To conclude this section, fractals may be unrepresentative. This is not a drawback but rather a very great strength from the viewpoint of education. If it is true that "math was never like that," it is also true that "this is more lifelike than any other branch of math."

8 The computer is the teacher's best friend in communicating the meaning of rigor

One passionate objection to the computer as the point of entry into real mathematics is the following: if the young replace solving traditional problems by computer games, they will never be able to understand the fundamental notion of mathematical rigor. This fear is based on an obvious chain of associations: the computer started as a tool of applied mathematics, applied mathematicians spurn rigor, the friend of my enemies is my enemy, therefore, the computer is the enemy of rigor.

With equal passion I think that the precise contrary is true: *rain or shine, the computer is rigor's only true friend.* True, a child can play forever with a ready-made program that draws Mandelbrot sets and never understand rigor, nor learn much of any value. But neither does the child who always does his mathematical homework with access to the teacher's answer book. On the contrary, the notion of rigor is of the essence for anyone who has been motivated to write a computer program—even a short one—from scratch.

When I was a student a non-rigorous proof did not scream *look out* at me and I soon realized that even my excellent teachers occasionally failed to notice clearcut errors in my papers. In the case of a computer program, on the contrary, being rigorous is not simply an esthetic requirement; in most cases, a non-rigorous program fails completely, and the slightest departure from absolute rigor makes it scream "Error!" at the programmer. No wonder that the birth of the computer was assisted by logicians and not mainstream mathematicians. (This topic is discussed in Mandelbrot 1993a.) It is true that, on occasion, a nonrigorous program generates meaningless typography or graphics, or—worse—sensible-looking output that happens to be wrong. But those rare examples only prove that programming requires no less care than does traditional proof.

Moreover, the computer programmer soon learns that a program that works on one computer, with its operating system, will not work on another. He will swear at the discrepancies, but I cannot imagine a better illustration of the changeability and arbitrariness of axiomatic systems.

Many other concepts used to be subtle and controversial before the computer made them become clear. Thus, computer graphics refreshes a distinction between fact and proof, one that many mathematicians prefer *not* to acknowledge but that Archimedes described wonderfully in these words: "Certain things first became clear to me by a mechanical method, although they had to be demonstrated by geometry afterwards because their investigation by the said mechanical method did not furnish an actual demonstration. But it is of course easier, when the method has previously given us some knowledge of the questions, to supply the proof than it is to find it without any previous knowledge. This is a reason why, in the case of the theorems that the volumes of a cone and a pyramid are one-third of the volumes of the cylinder and prism (respectively) having the same base and equal height, the proofs of which Eudoxus was the first to discover, no small share of the credit should be given to Democritus who was the first to state the fact, though without proof."

The first two sentences might easily have been written in our time by someone describing renascent experimental mathematics, but Archimedes lived from 287 to 212 BC, Democritus from 460 to 370 BC and Eudoxus from 408 to 355 BC. (Don't let your eyes glaze over at the names of these Ancient heroes. This chapter is almost over.)

When a child (and why not an adult?) becomes tired of seeing chaos and fractal games as white magic and draws up a list of observations he wants to really understand, he goes beyond playing the role of Democritus and on to playing the role of Eudoxus. Moreover, anyone's list of observations is bound to include several that are obviously mutually contradictory, stressing the need for a referee. Is there a better way of communicating another role for rigor and a role for further experimentation?

9 Conclusion

As was obvious all along, I am a working scientist fascinated by history and education, but totally ignorant of the literature of educational philosophy. I hope that some of my thoughts will be useful, but many must be commonplace or otherwise deserve to be credited to someone. One area where I claim no perverse originality is the historical assertions: they are documented facts, not anecdotes made up to justify a conclusion.

Now to conclude. The best is to quote myself and to ask once again: Is it extravagant to hope that, starting with this piece of mathematics called fractal geometry, we could help broaden the small band of those who see mathematics as essential? That band ought to include every educated citizen and therefore to have mathematics take its place among the liberal arts. A statement of hope is the best place to close.

Chapter 4

Mathematics and Society in the 20th Century[1]

Benoit B. Mandelbrot

Mathematics education and research are two separate crafts, but—for practical as well as intellectual reasons—it is best if they know each other. In particular, it is very important for mathematics educators to have a broad and balanced view of the way research mathematicians perceive their craft. They must realize that the perception has kept changing throughout history and never as sharply as in the 20th century. This chapter's goal is to recount a few highly significant features of the strife that came in the preceding hundred years. Mathematics ended that century in great spirit and in a state of great vigor, renewed collegiality and marvelous diversity.

But in the 1960s and 1970s, the representatives of the profession described the flow of 20th century mathematics as that of a single majestic river whose irresistible course was not touched by historical accident but had been preordained by inner logic. It necessarily proceeded inevitably and inexorably towards increasingly general, structural, or fundamental notions—which happened to be increasingly abstract. In the spirit of "the end of history," the descriptions never referred to the past or the messiness of Earth.

The majestic flow in question was unflinchingly understood to be leaving aside many people (including myself), and innumerable topics that concern either the foundations (logic) or the applications. We were told that much of what *looks* like mathematics is *not really* mathematics, even though the distinction may not be obvious to the outsider.

The position I am about to describe is starkly different. I believe and I hope to convince you that mathematics is *not* the conservatives' ivory tower. It is a very big house on a rolling terrain, with many doors, windows open to many horizons and bridges to many other houses.

It need not be the Queen of Court Etiquette in Science looking down on most of her subjects from an ivory tower up on a high hill. It deserves to be the beloved Queen of all the Scientists' Hearts, and of the Soul of Science, the only non-contrived link that could prevent various parts from scattering away from one another.

Compared to the conservative view of mathematics, mine is far broader and far more strongly linked to other human activities. It is also a more diverse and lively subject. In particular, it is attractive to persons who are not professional research mathematicians, a category that includes students and most teachers of mathematics. My strong opinions represent a minority view, but one that is increasingly widely shared and I have no doubt will prevail.

In any event, my interpretations and opinions are neither capricious nor based on idle rumor or anecdote, but on widely ranging reading, active and uninterrupted participation in events that occurred in the USA and France over fifty years, and reports by an uncle who was a prominent mathematician in Paris and Houston and participated in the immediately preceding thirty-five year period.

I see mathematical science as a very broad enterprise that shelters many diverse topics, ranging from the very concrete to the very abstract. This view is well represented by a simile I heard used by Hermann Weyl (1885–1955). He compared mathematics to the delta of a great river, one made of many streams: they may vary in their width and the speed of the flow through them; nevertheless, all are always a part of the system, and no individual stream is permanently the most important. This simile represented the mood of mathematics close to the year 1900—and also, for that matter, its mood near the year 1800. More importantly, mathematics has been changing so fast for a decade or so that I feel that Weyl's simile became applicable again in the year 2000.

But the resemblances between these snapshots taken centuries apart certainly do not imply that mathematics is unchanging, something outside ordinary history. In mathemat-

[1] Adapted from an invited address "What will remain of 20th century formal science" at the Europeäisches Forum 1992, held in Alpbach, Austria. This text remains self-contained and preserves some of its original flavor, in part by repeating some points that were already made elsewhere in this book but bear emphasis.

29

ics, as in every other aspect of human life, the 20th century gave us an example of something starkly different: a rocky history and continuing conflict. Mathematics was not ruled by its own determinism; it did not evolve separately from every other aspect of human knowing and feeling; it has on the contrary been profoundly affected by endless external vicissitudes.

The words *profoundly affected by* must not be misunderstood as meaning *enslaved by*. Of all the triumphs of humanity, the discovery and the development of mathematics is perhaps the greatest kind. A field's importance to the overall human experience is necessarily reflected by the role that internal logic has upon its development; nevertheless, strife has been present in mathematics since the Ancient Greeks. We shall see this when this story ends by mentioning the longstanding conflict between the traditions of Plato, the ideologue, and Archimedes, the experienced scientist. Like every individual human activity, mathematics very much participates in general history, politics, demography, and technology, and it is heavily influenced by the idiosyncrasies of a few key people. Let me give some examples from this century.

Around 1920, a group of Polish mathematicians collected around a very forceful man named Waclaw Sierpinski (1882–1969). They chose to concentrate on a field that was not practiced much in the reigning intellectual capitals, and founded a very abstract new branch often called *Polish mathematics*. They proudly proclaimed that their goal involved national politics: they did not want the newly reestablished Poland to become a mathematical satellite of Paris or Göttingen. I know that Providence is credited with working in mysterious ways. Yet, would anyone claim that Polish nationalism after more than a century of partition had anything to do with the historical determinism of mathematics? Polish mathematics became an important force pushing towards abstraction at all cost. Yet, by a bitter irony, some of the notions it originated failed to become important in mathematics, but eventually became important to physics—through fractal geometry.

My second example concerns Godfrey Harold Hardy (1877–1947), a strong person as well as a strong and highly inventive mind. The Poles had no strong native physics to contend with, but British mathematics of Hardy's youth was dominated by a form of mathematical physics that was extraordinarily effective (the Heaviside Calculus differentiated discontinuous functions!) but had little concern with continental rigor. During World War I, Hardy was an outspoken pacifist who recoiled from the practical uses of this old British mathematics. During another War, he wrote (Hardy (1940)), an impassioned account of his ideal of pure mathematics. For him, good mathematics could have no bad application—for the simple reason that it could have *no* application of any sort. By another bitter irony, his best example of total inapplicability turned out, in due time, to be essential to a problem he would have loathed: cryptography.

A three-page review of Hardy (1940) in the famous weekly *Nature* by the Nobel-winning chemist Frederick Soddy begins "This is a slight book. From such cloistral clowning the world sickens ... 'Imaginary' universes are so much more beautiful than this stupidly constructed 'real' one, according to the author ... Most scientists, however, still believe that ... the real universe ... is not stupidly constructed." But nothing can break the appeal of a tract that discriminates between the good and the bad without hesitation. Hardy's book remains in print and continues to this day to enchant some of the young. But would anyone claim that Hardy's militant anti-nationalism had anything to do with the historical determinism of mathematics?

From ideology, let us move on to demography. The 1910s were very cruel to French mathematics. First, Henri Poincaré (1854–1912) died prematurely on the operating table, then millions of young people died in trench warfare, and finally—perhaps worst of all—millions returned broken in health or spirit to a country that did not dare make heavy demands on them. As a result, the young postwar French mathematicians of the 1920s found that the only available teachers were men who had already been ill or old in 1914 and so did not go to war. Some have written movingly about the hardship of training without the usual parental supervision from slightly older advisors, and (as may have been expected) this hardship contributed to the emergence of several very strong personalities. In any event, the France of the late 1920s and the 1930s gave rise to an extremist movement calling itself Bourbaki. But would anyone claim that a demographic unbalance in a country with a long and glorious mathematical tradition has anything to do with the historical determinism of mathematics?

André Weil (1906–1994), now acknowledged as the mind behind Bourbaki, observed late in life that in his prime years, mathematics was little influenced by physics. Was that a natural feature of the preordained development of mathematics? Or could it be that Weil's views were set even before a visit to Göttingen in the 1920s? David Hilbert's dream Mathematics Institute there had three parts: a very pure one that Weil worshipped, one on numerical methods and one on mathematical physics. In the latter part, Max Born and Werner Heisenberg were in the process of creating quantum mechanics—but Weil apparently did not notice.

From demography, let us move to another form of ideology. Soviet anti-semitism treated Jewish mathematicians harshly; Jewish physicists, less so. Hence a number of very gifted mathematicians transferred to physics institutes, where they were welcome. Their move contributed greatly to the formation of the current very rigorous form of mathematical physics. Would anyone claim that Soviet ethnic politics have anything to do with the historical determinism of mathematics?

No one would claim that the specific historical determinism of mathematics only reflected the intellectual moods and fashions that rule society at large. But it happens that a very unusual mood prevailed early in this century, particularly in the 1920s. One especially visible and durable effect was the invention of the International Style in architecture, with its heavy emphasis on structure. In Finland, the very unusual small country where this style was born, modern architecture merged smoothly into what came before it, without discontinuity and without heavy dogmatism. But modern architecture became dogmatic in Germany with the Bauhaus and in France with Le Corbusier (1887–1965). The latter built few houses but made many sketches (for example, his proposed ideal improvement of Paris evokes the worst present suburbs of Moscow). When I was young, Le Corbusier was billed as a great intellect to whom modern architecture owed its intellectual legitimacy. Indeed, he wrote a great deal, but I find little in his writings beyond sophomoric trash. It may be that Bauhaus was useful, even commercially inevitable at a certain stage of the technology and economics of raising large buildings, but no one ever convinced me that they were an inevitable intellectual wave of the future.

Think also of physics. Having confirmed existence of the atom in the 1900s, it went on to focus increasingly on the search for the most fundamental structural components of matter, increasingly tiny ones. Biology took this path later.

How was mathematics affected by the above-mentioned politics, demographics, and general intellectual moods? I view them all as responsible for the fact that near the middle of our century mathematics behaved in ways totally at variance with its mood today and its mood in 1900 or 1800.

This atypical mathematics is conveniently denoted by the name it took in France, but the current that gave rise to Bourbaki also affected many countries other than Britain, France and Poland. It strongly affected the USA, with a little-known wrinkle. One might have expected a brash new industrial giant to favor applications, but in terms of mathematical research the precise contrary was true. In Europe, the 19th century had created wide-ranging establishments against which Bourbaki could revolt. In the USA, before the arrival of refugees from Stalin and Hitler, research mathematics was dominated by aristocrats and anarchists, hence was very pure (as well as outstanding on its terms). Bourbaki did not reach the outlying countries Sweden and Finland, and there were strong counteracting forces in Germany and Russia. In the 1960s, when Bourbaki was its strongest, it benefited from another extraneous event: Sputnik created a period of unprecedented economic growth in Academia, with minimal social pressure on the sciences, and greatly increased the number of math PhDs, including many Bourbaki products. The math departments' balance was overwhelmed by them.

To sum up, Bourbaki found roots by selecting one of the many components of the mathematics of 1875–1925, gathered strength during the second quarter of our century (the period to which the above examples refer), and took power around 1950. During the third quarter of the century it ex-

erted an extraordinary degree of control. There was no disorder in mathematics, but the field was narrowed down to a truly extraordinary extent. At one time it seemed to reduce to little more than algebraic topology; at a later time, to number theory and algebraic geometry. These are extremely important fields, to be sure, but concentration on a single field was quite contrary to the historical tradition that I have already mentioned and that had led Hermann Weyl to the image of the delta of the Nile. Mathematics seemed to have reduced itself to basically a single stream at any given time. This happened to be the cliché description that Herman Weyl (in a contrasting image) applied to physics.

The Bourbaki, as has already been implied, never paid attention to the historical accidents that contributed to their birth; they felt themselves to be the necessary and inevitable response to the call of history. Today, however, this call seems forgotten, and there is wide consensus that, like new math, "Bourbaki is dead."

Who killed Bourbaki? Throughout its heyday, my friend Mark Kac (1914–84) and many other open-minded mathematicians argued, in vehement speeches and articles, that Bourbaki had misread mere accidents for the arrow of history. But such negative criticism invariably lacked bite, and it had no effect. My own partisan opinion is that Bourbaki's fate was typical of many ideologies outside science. The founders could only insure their immediate succession; gradually, the ideological fervor weakened and the movement continued largely by force of habit. The resulting weakening was gradual and not obvious. But everyone noticed when the movement was knocked down by yet another event that had nothing to do with the historical determinism in mathematics. This event was something I view as a return to sanity, namely the rebirth of experimental mathematics that followed (slowly, as we shall see) the advent of the modern computer.

From where did the computer come? From the mathematical sciences understood in a broad sense. What relation is there between the advent of the computer and mathematics as narrowly reinterpreted by Bourbaki? None whatsoever. The computer arose from the convergence of two fields that surely belong to mathematics but were spurned by Bourbaki, namely, logic and differential equations. We all know that one must never rewrite history as it might have proceeded if two crucial events had chanced to occur in the reverse order. But in this instance the temptation is strong to air the following conviction. Had an earlier arrival of the computer saved experimental mathematics from falling into a century of decline, Bourbaki might have never seemed to anyone to be an unavoidable development. Let me elaborate on the computer's roots.

Surprisingly, while **Foundations of Analysis** was (for a while) the overall title of their treatises, the Bourbaki had only contempt for the logical foundations of mathematics, as in the work of Kurt Gödel (1906–78) and Alan Turing (1912–54). In the 1930s, Turing had phrased his model of a logical system in terms of an idealized computer. His *Turing machine* had a very great influence on the thinking of those who developed the actual hardware.

However, the man who made the computer into a reality was John von Neumann (1903–57). He was not only a mathematician, but also a physicist and an economist, and his great breadth of interests came to include a passion to find ways to predict the weather.

Thus the computer was born in the 1940s from a strange combination of abstract logic and the desire to control Nature. Eventually, the computer changed mathematics in a very profound fashion. But for a very long time, core mathematicians felt totally unconcerned, and viewed it with revulsion. Because of his work on the computer, von Neumann ceased to be accepted as a mathematician and in 1955 he decided to resign from the Princeton Institute for Advanced Study. (He died before his planned move to California.)

The year 1955 was also the date of publication of Fermi, Pasta & Ulam (1955), a text that appeared only in a Los Alamos report but was widely read and viewed as an early masterpiece of experimental mathematics before it was actually printed in Fermi (1965) (pp 977–988) and then in Ulam (1974) (pp 490–501). A comment by Stanislaw Ulam (1909–1984) informs us that the initiative for using the computer to assist mathematical research had come from Enrico Fermi (1901–1954), who was of course a physicist, not a mathematician. And Ulam asserts that "Mathematics is not really an observational science and not even an experimental one. Nevertheless, computations were useful in establishing some rather curious facts about simple mathematical objects." Surprised, I reached for a more positive statement in the autobiography, Ulam (1976), but found nothing worth quoting.

How did experimental mathematics fare during the 25 years after 1955? That period happens to end in the year of publication of Mandelbrot (1980), and coincided with the heyday of Bourbaki. In a near-perfect first approximation, it saw *no experimental mathematics at all*. Not only was the lead of von Neumann and Fermi not followed by mathematicians, but their disinterest for the computer was carefully considered, not caused by ignorance. For example, when I was new at IBM, which I joined in 1958, opportunities to use computers were knowingly and systematically offered to every mathematician with a good name who could be coaxed into the building. Not one of them paid attention to the offer. Interest in experiment did not spread to at least some mathematicians until my work started attracting wide attention.

In understanding the process of discovery, the slowness of the acceptance of the computer brings up forcibly a very old issue: the respective contributions of the tool and of its user. Galileo wrote a whole book complaining bitterly about those who belittled him by claiming that his discovery of sunspots was only due to his having lived during the telescope revolu-

tion. In fact, telescopes were widespread but useless before one reached Galileo's steady hand and good eye. For contrast, consider the chapter of mathematics called the global theory of iteration of rational functions, to which the Mandelbrot set belongs. Pierre Fatou (1878–1929) and Gaston Julia (1893–1978) are—quite rightly—praised for developing this chapter, and no one would dream of belittling their contributions as being due to their having lived during the age of Paul Montel (1876–1975). Montel was the mathematician who, in 1912, discovered Fatou's and Julia's key tool, called the normal families of functions. Soon afterwards he was called into the Army, leaving behind Fatou (who was a cripple) and Julia (who had come back from the trenches as a wounded war hero). After World War I, Montel looked after the theory of iteration as *his* baby. Today, in the noise that accompanies changes in mathematics, those who use the computer are treated like a Galileo and not like a Montel. That is, critics are found to belittle their work as solely due to their living in the computer age. If it were so, experimental mathematics would have thrived after von Neumann and Fermi; the preceding remarks show that it did not.

Let me summarize, make a general comment, and conclude. One cannot disregard the lessons of history, contrary to the belief of those who argued that the pure mathematics of the mid-20th century was preordained by destiny. Its birth in the 1920s was influenced by Polish and English ideology, a demographic catastrophe in France, and the general intellectual mood of the day; its success was hastened by a long spurt of economic growth, and its demise was hastened by a mere technological development. None of these events was influenced by mathematics, none was preordained, and none acted immediately. In any event, von Neumann's and Fermi's lead was not followed by other mathematicians.

To conclude, "What will remain of 20th century mathematics?" There can be no short and truthful answer, because at this point of its history, mathematics is in healthy and constructive turmoil. Once again, the Bourbaki utopia flourished when every science was experiencing unprecedented growth and minimal social pressure. It seemed that any would-be peer group could organize itself and prosper with no hindrance from other, equally self-interested peer groups. But today the sciences face scarcity and strong pressure to justify both their size and their goals, and everyone bemoans the absence of generalists capable of representing more than a few groups. How the effects of the resulting intractabilities and pressures will combine with the internal logic of mathematics, of the computer and also of today's mathematical physics— that thriving no man's land between theorem proving and observation of nature—is simply beyond prediction. Fortunately for the teachers of mathematics, they are not asked to predict, but it is best for them to know the past, if only to avoid being drawn to repeating its deep errors.

Classroom Experiences

Chapter 5

Teaching Fractals and Dynamical Systems at The Hotchkiss School

Melkana Brakalova and David Coughlin

1 Introduction

This is a report on the development of a Fractals and Dynamical Systems component within the math curriculum at The Hotchkiss School, a private preparatory school in northwestern Connecticut. We will briefly describe the genesis of the relevant courses and then present in some detail various aspects of the teaching methods, problems, student projects, and student reactions to the courses. We hope that some of what we have learned will be of help to others involved in similar curricular experiments at the high school and college level. Certainly, at The Hotchkiss School courses on fractals and dynamical systems have entered the normal curriculum—the courses are taught within our standard weekly schedule, four times weekly for 45 minutes—and have been, in our judgment, quite successful. Though we will no doubt continue to refine our approach to the teaching of the subject, at least in this school, fractals and dynamical systems are here to stay.

2 History and structure of the course

2.1 Development of the course

In the first semester of the 1992–93 academic year two members of the Hotchkiss Mathematics Department (D. Coughlin and W. Gaynor) offered an independent study course to three seniors who wanted to continue mathematics beyond the BC Calculus level. The offering was inspired by a lecture of Professor Devaney during the CAIS–Math Forum at The Hotchkiss School in April 1990. The instruction was somewhat loose and experimental. At the end of the term these students responded to a final term project with enthusiasm and excitement, and made a multimedia presentation to which the school community was invited. They learned about the Mandelbrot and Julia sets from the book Devaney (1990), and were able to articulate and share these ideas with their peers. These presentations were so successful that they became a model for end-of-course projects in succeeding years.

Subsequently, the first author designed and offered a new course, entitled *Topics in College Math: Fractals and Dynamical Systems*. Assisted by the second author, she taught this course in the second semester of the 1993–94 academic year. The course was taken by four mathematically advanced and gifted students who had already completed a semester of Multivariable Differential Calculus. This new course was designed to combine the excitement of computer imaging and a familiarity with deep mathematical thinking developed during the students' calculus classes. The main text for this course was Devaney (1992). Additional materials included videotapes, computer software, and graphing calculators. The course concluded with end-of-term projects and multimedia presentations attended by the entire Math Department and other members of the school community.

Our experience was sufficiently encouraging that we offered a similar course, in the second semester of 1994–95, to a larger group of students (7) who we judged had sufficient math skills and interest. The enrollment requirement was a semester of Differential Calculus. The course, *Fractals and Dynamical Systems*, was taught by both authors. The required texts were Devaney (1992) and Gleick (1987).

In this chapter we present mainly the material taught during the 1994–1995 year, much of which is identical to that taught in 1993–94. In addition, a description of the students projects from both years is included. Since every year brings students with diverse backgrounds and abilities, we adjust the course to the students' levels and interests. Thus the course is changing and evolving in new directions and our intention is to respond in the best possible way to such developments.

2.2 Content of the courses

Here we discuss the following topics, as they were introduced in the two courses. The numbering refers to the sections of this chapter where each topic is discussed.

3. Fractals
4. Discrete Dynamical Systems
    4.1. Introduction
    4.2. The linear function
    4.3. Quadratic functions
    4.4. The Orbit diagram
    4.5. The Bifurcation diagram
    4.6. Doubling and Tent Functions
    4.7. The Feigenbaum Constant
5. Complex Numbers and Complex Functions
6. The Mandelbrot and Julia sets
7. Newton's Method

2.3 Teaching methods

The 1993–1994 course was taught to a small group of very mathematically advanced students whose backgrounds made it possible to conduct the course in a relaxed and informal seminar format. For example, the students were asked to participate actively in the presentation of new material. This was complemented by discussions of assigned homework problems and additional presentations on research projects, some of which involved computer work. The rather few quizzes and tests included set problems but also required students to address broader theoretical aspects of the material.

During the 1994–95 course we largely preserved the relaxed and informal format, but with less emphasis on student involvement in the presentation of new material. This shift reflects our judgment that the 1994–95 students were on the whole somewhat less advanced than the previous group. Rather, the new material was presented in a lecture format or through worksheets as in-class activities. Still other classes were devoted primarily to student presentations of the solutions to homework problems. At several points in the course, students were given mini-projects, and their presentations occupied the entire class time. A number of video tapes on these topics are available; some were shown for either a portion of or for the entire period of several classes. The videotapes were genuinely instructive and enjoyed by the students.

The end of course projects were either assigned by an instructor or chosen by the students. Some students worked individually, others in pairs. Following the model of our first year, the students were asked to make oral presentations of these projects—to a general student audience—and also to submit a written report. The oral presentations, usually accompanied with overhead transparencies, a calculator overhead view screen, or a computer, proved to be most valuable. The process encouraged students to so familiarize themselves with the material that they were able to express themselves in a fluent and concise manner. This was a new experience for many students, and it helped publicize the course and generate interest among other students.

3 Introduction to fractals

So far, our experience in teaching fractals has been devoted to the introduction of self-similar geometric shapes such as the Koch Snowflake, Sierpinski Triangle, Sierpinski Carpet, Sierpinski Tetrahedron, Cantor Middle Thirds Set, and others. We emphasized the iterative nature of the construction process of these figures and how such a process creates self-similarity. Using self-similarity as our foundation, we introduced the definition of fractal dimension as the ratio of the log of the number of self-similar pieces to the log of the magnification factor. We used segments, squares and cubes to demonstrate how this definition works in the case of the usual 1, 2, and 3 dimensions. Students sometimes were asked to create their own fractals and to compute their dimensions. The fractals were defined using an original shape to which an iterative process was applied. On some occasions we have used Fractal Exploration Kit software, Choate (s1990), to create linear and coastline fractals.

In the study of fractals shapes, we computed perimeters, areas, and volumes, (where appropriate) using sums of geometric series. Students were amazed to discover how an infinite boundary may enclose a finite area or volume.

At the beginning of the first year, we gave this exercise, from Peitgen, Jurgens, & Saupe (1991).

The Sierpinski Triangle

The top of Figure 1 illustrates the procedure for generating a fractal called the Sierpinski Triangle. The starting figure is an equilateral triangle with sides of length 1. The next step consists of removing the white triangle in the middle with sides of length 1/2, the third step consists of removing the three white triangles with sides of length 1/4 and so on *ad infinitum*. The set remaining after the removal of all triangles in this process is called the Sierpinski Triangle. Sudents are asked to find the area of all triangles removed in the infinite process and compare the result to the area of the original triangle, then to discuss their findings.

A similar project was assigned on the area and the perimeter of the Koch snowflake. See the bottom of Figure 1. The realization that the Sierpinski triangle has no area and that the Koch snowflake occupies a finite area but has an infinite perimeter helped to illustrate the counterintuitive nature of some simple fractal objects. More complicated fractal shapes,

Figure 1: Top: Three steps in the creation of the Sierpinski triangle. Bottom: Three steps in the creation of the Koch snowflake.

including Julia sets and the Mandelbrot set, are addressed later in the chapter.

4 Discrete dynamical systems

4.1 Introduction

Students were introduced to dynamical systems as defined by two components: an *initial state* (or *initial value* or *seed*) and a *dynamic*. In our case the dynamic consisted of iterating a function, where the current output of the function becomes the next input. When this process is repeated, a sequence of numbers (or symbols) is produced. This sequence is called the *orbit* of the seed x_0. There are various ways to write the orbit, for example as $x_0, f(x_0), f^2(x_0), \ldots, f^n(x_0), \ldots,$ where f^n does not mean the power but rather the nth composition of f. To examine the orbits of different seeds under iteration of different functions, we used separate features of the *TI-82* graphing calculator—the *ANS* key and the *WEB* diagram. We also did *graphical analysis* by pencil. The latter consists of drawing the vertical line from (x_0, x_0) to $(x_0, f(x_0))$, then drawing the horizontal line from $(x_0, f(x_0))$ to $(f(x_0), f(x_0))$. Repeat to $(f^2(x_0), f^2(x_0))$, $(f^3(x_0), f^3(x_0))$, and so on. An example involving the logistic function $Kx(1 - x)$ for the case $K = 2$ was used to study the orbit of 0.5. Using the *GRAPH* key one gets the display of the graph of the function $y = 2x(1 - x)$ on the viewing rectangle together with the graph of the line $y = x$. The *TRACE* key locates the seed and by pressing the right arrow key one gets the orbit of the chosen seed value. This exercise was appreciated and enjoyed by the students.

Faster methods for obtaining longer displays of an orbit (thousands of iterates) are available on computers. One can use a spreadsheet or software such as Georges, Johnson & Devaney (s1992). The latter displays the orbit values, the time series and the web diagram. We have used this software to predict the behavior of the orbit in more complicated cases when a pattern appears after a certain number of initial iterates.

In addition, at this early stage students were introduced to applications such as the Babylonian method for finding square roots, and also some problems on population growth (see Chapters 1 and 2 from Devaney (1992)).

We used specific examples to introduce different types of orbits: orbits consisting of fixed points, orbits converging to fixed points, eventually fixed points, orbits that tend to infinity, periodic orbits, eventually periodic orbits, orbits that converge to periodic points, and chaotic orbits. Here is an example of a worksheet we used in class.

Types of Orbits

A. *Fixed Point.* The point x_0 is a fixed point of $F(x)$ if and only if $F(x_0) = x_0$.

Fixed points are found by solving the equation $F(x) = x$, if possible. Graphically, fixed points can be found at the intersection(s) of the graphs of $y = F(x)$ and $y = x$.

Example. $F(x) = x^2, x_0 = 1$.

There are several kinds of fixed points depending on the behavior of the orbits of nearby seeds.

Exercise: What are the fixed points for $y = \sqrt{x}$? Compare the behavior of orbits of points near each of them. What is the fate of the orbit of $x_0 = 1.1$, of $x_0 = 0.1$?

B. *Orbits that converge.*
Example. $F(x) = \sqrt{x}$; $x_0 = 24$.
Example. $C(x) = \cos x$; $x_0 = 7$.

C. *Eventually Fixed Orbits.*
Example. $F(x) = |2x - 1|$; $x_0 = \frac{1}{2}$.
Is x_0 a fixed point or is it a matter of convergence? Does F have any other fixed points?

D. *Orbits that tend to infinity.*
Example.

$$F(x) = x^2; \quad x_0 = 1.0001.$$

Then $\lim_{n \to \infty} F^n(x) = \infty$.
Note the difference between the example in part A, where the seed was 1, and this example, where the seed is 1.0001. Such a distinction in the orbit behavior of nearby points is an example of *sensitive dependence on initial conditions*.

E. *Periodic orbits.*
The orbit of x_0 is periodic under F if $F^n(x_0) = x_0$ for some $n > 1$. The smallest value of n is called the prime period of the orbit. Usually if the prime period is 4, for example, we say "x_0 has period 4" or "x_0 belongs to a 4-cycle".
Example. $F(x) = |x - 1|$; $x_0 = 0.9$.

F. *Eventually periodic orbits.*
Example. $F(x) = x^2 - 1$; $x_0 = 1$.

G. *Chaotic orbits.*
Example. $F(x) = x^2 - 2$; $x_0 = 0.25$.

4.2   The linear function

We used graphical analysis to study the effect of the slope of a linear function on the behavior of different orbits. These observations were generalized in a project where students were asked to predict the behavior of any fixed seed x_0 under a general linear function $y = ax + b$ (a not equal to 0). Proofs of their findings serve as preparation for the study of the orbit behavior near attracting and repelling fixed or periodic points.

4.3   Quadratic functions

By experiment we studied the fate of the orbits of seeds for different values of the parameter of the quadratic and logistic functions. For example we examined orbit behavior using the web diagram on the *TI-82* graphing calculator, the software Georges, Devaney, & Johnson (s1992), and spreadsheets. We designed examples of quadratic functions and seeds for which the eventual behavior of the orbits was independent of the seed. Students observed some patterns emerging. These patterns were addressed in detail during the study of the orbit and the bifurcation diagrams.

4.4   The orbit diagram

We introduced the orbit diagram of the quadratic and logistic functions through computer projects following the main idea of the experiment described in Devaney (1992), Chapter 6.4. The next two projects were assigned to different groups of students.

Orbit Diagram
Project 1

1. Purpose. To investigate the behavior of the orbit of $F(x) = \lambda x(1 - x)$, as the parameter λ increases from 0 to 4.

2. Procedure. Using Devaney's software or a spreadsheet, each student finds the first 2000 iterates of the seed $x_0 = 0.5$ for the assigned values of the parameter λ. For each parameter value, students decide the type of orbit of x_0. If the orbit converges to a fixed point or to a cycle, students list the periodic points to two decimal places. If the orbit does not seem to converge to anything, students list the last eight iterates from the spreadsheet. On a printout of the last 20–30 iterates, students list the points they are going to graph. They write brief explanations of the asymptotic behavior observed.

    On the provided graph paper students very carefully plot all the points with the x-coordinate equal to one of the parameter values and the y-coordinate equal to one of the corresponding points of the periodic cycle to which 0.5 converges. If the seed 0.5 does not converge to any periodic cycle, students used the last eight iterates from the spreadsheet for y-coordinates.

3. In class exercise. On transparencies we collect the data from each student and assemble a composite graph of all orbits studied by overlaying the transparencies. The picture we get is called an orbit diagram.

Parameter Values

Set 1: 0, 0.5, 1, 1.5, 2, 2.5, 3, 3.1, 3.2, 3.3, 3.4, 3.45, 3.6, 3.7, 3.8, 3.9, 4.

Set 2. 0.1, 0.6, 1.1, 1.6, 2.1, 2.6, 3.05, 3.22, 3.32, 3.42, 3.46, 3.62, 3.72, 3.82, 3.92.

Set 3. 0.2, 0.7, 1.2, 1.7, 2.2, 2.7, 3.07, 3.24, 3.33, 3.44, 3.47, 3.64, 3.73, 3.83, 3.93.

Set 4. 0.3, 0.8, 1.3, 1.8, 2.3, 2.8, 3.09, 3.25, 3.34, 3.46, 3.48, 3.65, 3.74, 3.84, 3.94.

Set 5. 0.4, 0.9, 1.4, 1.9, 2.4, 2.9, 3.15, 3.27, 3.35, 3.47, 3.49, 3.67, 3.75, 3.85, 3.95.

Set 6. 0.45, 0.95, 1.45, 1.95, 2.45, 2.95, 3.17, 3.28, 3.36, 3.48, 3.49, 3.67, 3.75, 3.85, 3.96.

Project 2

This is the idea of Project 1, applied to the quadratic function $F(x) = x^2 + c$, where the parameter c decreaseses from 0 to -2 and the seed is 0.

An important application of the logistic function is as a model for population growth. The variable x represents the fraction of the maximum population the environment can support, and the reproduction pattern of the population determines the coefficient λ. Such a model was applied by the biologist Robert May (1976) who studied the behavior of populations with different coefficients. His investigations led to the creation of the orbit diagram, and we found it valuable to refer to his research as described in Gleick (1987). We assigned a reading from that book's chapter "Life's Ups and Downs," and reinforced some of the main facts and ideas with a quiz:

1. Who is Robert May? (background, interest, education, time of research)

2. What is the ecological question that May studied? What model did he use to investigate it? What most intrigued him when studying this model? How did May define the point of accumulation?

3. What is the important gap that May filled with his studies (e.g., did mathematicians or biologists study the same phenomena before?) and why was he in an advantageous position to do so?

4. Discuss the outline of the bifurcation diagram as May first saw it.

5. In few words summarize the significance of the discovery of the bifurcation diagram.

Our experience showed us that students are not used to dealing with such assignments in math classes and that a reading should be preceded by a questionnaire to focus their attention on some of the important facts and ideas.

We also handed out some articles from the *Mathematical Intelligencer* discussing Gleick's book *Chaos* (Devlin (1989), Douglas (1989), Franks (1989a), Franks (1989b), Hirsch (1989), Gleick (1989)). That mathematical ideas can be controversial topics in modern times was something that most students appreciated.

4.5   Calculus of fixed points and bifurcation

The reason for the prerequisite of one semester of differential calculus became apparent when we introduced the notion of attracting and repelling fixed points and cycles. These properties can be easily expressed in the language of differential calculus. Furthermore, some important theorems of calculus, such as the Mean Value Theorem, can be used to prove observed patterns in the dynamic behavior. An excellent treatment of this topic is provided in Devaney (1992). We largely followed his approach and used many of the exercises provided.

We examined attracting, neutral, and repelling fixed points based on the absolute value of the derivative at the fixed point: is it greater than, equal to, or less than one? This is a place where one can link the earlier findings about the slope of the linear functions to the value of the derivative. We introduced and proved the Attracting and Repelling Fixed Point theorems in Devaney (1992). The ease with which the proofs were understood depended on the level of mathematical preparation of the students. Those who had completed BC Calculus did better than those who had not.

When studying periodic points we emphasized two properties: first, that a fixed point of a function F is also a fixed point of any iterate of F. Second, that a periodic point of period n is attracting (or repelling) if it is an attracting (or repelling) fixed point of the nth iterate of F. For the latter, we introduced the Chain Rule Along a Cycle.

We spent some time investigating the behavior of orbits nearby a neutral fixed point. Experiment and observation were effective here. Students were given a sample of polynomials and transcendental functions with neutral fixed points. They had to link the value of the first, second and third derivatives at the fixed point to the orbit behavior, and then make a conjecture. The experiment was quite successful and the conjectures were very close to the Neutral Fixed Point Theorem.

During the second year of teaching *Dynamical Systems and Fractals* we spent a considerable amount of time on the construction of the first few branches of the bifurcation diagram of different second degree polynomials. We also showed that one can obtain one bifurcation diagram from another using an appropriate transformation.

We began with the quadratic function $Q_c(x) = x^2 + c$ and found expressions for the two fixed points as functions of the parameter. Then we determined for which parameter values these points exist and are attracting, repelling or neutral. Continuing with the same quadratic function, saddle-node (or tangent) and period-doubling bifurcations were defined and discussed. Long division was used to determine the expressions for the periodic points of period 2. With these expressions it was possible to find the exact parameter value where the first and second period-doubling bifurcations occur and therefore to draw a portion of the bifurcation diagram. The same exercise was repeated with the logistic function and a take home test involved the same questions for another second degree polynomial.

The transparencies, containing the data from the previously described experiment on orbit analysis of the quadratic and logistic functions, were superimposed over the graphic images (produced on the viewing screen of the graphing calculator) of the corresponding bifurcation diagrams and an astonishing match (the scales were almost the same) was observed.

We concluded this unit with this take-home test.

Consider the function, $F_c(x) = x^2 + x + c$, where c is a parameter.

1. Find expressions for p_+ and p_-, the fixed points of F. For what values of c does F have fixed points?

2. For what values of c is p_+ attracting? For what values of c is p_- attracting? Show your justification clearly.

3. At what value of c does a saddle-node bifurcation occur? Explain your answer by demonstrating your knowledge of the definition of a saddle-node bifurcation.

4. Find the period-2 points in terms of c. (Hint: $(a + b + c)^2 = a^2 + b^2 + c^2 + 2ab + 2ac + 2bc$.)

5. For what values of c, if any, is the 2-cycle attracting? Show your work using the Chain Rule Along a Cycle.

6. At what value of c does a period-doubling bifurcation occur? Justify your answer demonstrating your knowledge of the definition of a period-doubling bifurcation.

7. At what value of c does the next period-doubling bifurcation occur?

8. Draw as much as you can of the bifurcation and orbit diagrams for F. Label axes and indicate units on both axes.

4.6   The doubling and tent functions

We introduced the doubling and tent functions and their properties mainly following the text and problems in Devaney (1992), pp 24 and 28. Here the doubling function is defined in the unit interval as $D(x) = 2x \pmod 1$, and the tent function $T(x) = -|3x - 3/2| + 3/2$ is defined over the real line. We related the fate of the orbit of a seed to its binary expansion (in the case of the doubling function) or its ternary expansion (in the case of the tent function).

Here is an example of a take-home project.

The doubling function and chaotic behavior

1. Find the decimal number A that has binary expansion $0.(1101)^-$, that is, the sequence 1101 repeats forever. Find the first five iterates of A under the doubling function using

    (a) its binary expansion,

    (b) its decimal presentation.

    Determine if the orbit of A is periodic, fixed, eventually periodic, eventually fixed or it exhibits chaotic behavior.

2. Let A be any rational number. What types of orbits may A have and why?

3. Suggest a number that you think will have a chaotic orbit under the doubling function. Check if this orbit is chaotic by looking at the first 100 iterates on your graphing calculator. Are there any discrepancies between what you expect and what your calculator suggests?

Here is a class exercise to introduce some of the properties of the tent function.

Tent Function: In-class exercise.

1. Find an expression that will give you $T^{10}(-0.2)$ and $T^{10}(1.2)$. Use the answer key on the *TI-82* to confirm these results.

2. Find a general formula for $T^n(x)$ for $x < 0$ or $x > 1$.

3. What happens to $T^n(x)$ as n approaches infinity when $x < 0$ or $x > 1$?

4. Summarize these observations in words about the eventual behavior of the orbit of a seed outside the closed unit interval.

5. Predict the behavior of the orbits of seeds in the middle third interval under $T^n(x)$ for $n = 1, 2, 3$ and 4.

6. Predict the behavior of the orbits of seeds in the middle third interval under $T^n(x)$, for any n.

The superimposed images of the tent function and its first few iterates on the graphing calculator provided an insight into the fate of the orbits of seeds lying in the middle thirds intervals. We linked these graphs to the Cantor Middle Thirds set and discussed its uncountability using binary and ternary expansions of points in the unit interval.

4.7 Feigenbaum's constant

We asked the students to determine experimentally the approximate values of the parameter for which period doubling bifurcation occurs. For this experiment they used the orbit diagrams available on Georges, Johnson, & Devaney (s1992) for the logistic, quadratic, and other functions. After determining these parameters the students were asked to find the

ratios of the lengths of the consecutive intervals for each of the functions. In addition, a reading from the chapter "Universality" of Gleick (1987) was assigned. We asked our students to compare their results with those discussed in the reading and to write a short paragraph on their findings. Because of the delicate nature of this experiment some of the results were not close to the reliable approximate values.

The computation of Feigenbaum's constant was also assigned as a final project. It is described in Section 8.2.1 and it involves the use of both super attractive parameters and Newton's Method.

5   Complex numbers and complex functions

We started our exploration of complex functions by defining equivalent ways (polar, Cartesian, binomial) for presenting complex numbers and of performing algebraic operations on them. The modulus and the argument were used to represent analytically simple geometric shapes such as rays, lines, circles, ellipses, and interior angles. We also discussed complex functions including polynomials, e^z, $\cos z$, $\sin z$, $\tan z$ and their formal derivatives. We approached iterations by first studying the dynamics of $F(z) = az + b$. The two problems shown here are part of a quiz on this material.

Complex numbers and the linear function

1. For each of the following three functions

    1.1 $L(z) = \dfrac{i}{2}z,$

    1.2 $L(z) = \dfrac{1}{2}(1 + \sqrt{3}i)z,$ and

    1.3 $L(z) = (-1 + \sqrt{3}i)z:$

    (a) Determine and plot the first five iterates of the orbit of $z_0 = 1$.

    (b) Predict the eventual behavior of the orbit. (Be as specific as possible, i.e., is the orbit periodic, dense, does it tend to infinity or zero?)

2. Find the fixed points of $F(z) = z^3 + (1 + i)z$.

After studying the linear function, we continued to the quadratic function and the definition of Julia sets and the Mandelbrot set.

6   The Mandelbrot set and Julia sets

We introduced the Mandelbrot set and Julia sets by a combination of texts, software and videos, including Devaney (1992), Devaney (v1989), and Devaney (v1990).

During the second course, offered in 1994–1995, we greatly benefited from an all-school presentation by Dr. Devaney himself, who lectured on the Mandelbrot set and the corresponding Julia sets. The lecture was accompanied by a short movie of continuously changing images of Julia sets as the value of the parameter changes inside and outside the Mandelbrot set. This movie inspired some of our students who designed their own project on the study of higher degree Mandelbrot set and corresponding Julia sets, described in Section 8.2.3. The students were also shown Peitgen, Jurgens, Saupe and Zahlten (v1990). This exceptionally successful educational movie linked together many components of the course and nicely introduced the next topic, Newton's method.

7   Newton's method

The unit on Newton's Method seemed especially successful and interesting to the students. It combined traditional mathematics, calculator use, and intriguing computer images. Several students chose to expand on our discussion of this topic in their end-of-term projects. We supplemented the section on Newton's Method in Devaney (1992) with Parris (1990), the appendix to Berkey & Blanchard (1992), and a chapter from another book on chaos, Peak & Frame (1994).

We began our discussion of Newton's Method by reviewing Newton's Method for finding the real roots of equations. Students were asked to find a real root, to several decimal places, for $F(x) = x^3 - 3x + 1$, using Newton's Method and a calculator. We then repeated the exercise using the *ANS* key on the *TI-82* calculator, emphasizing the iterative nature of this method. Finally, we introduced the Newton Iteration Function, $N(x)$, associated with any function $F(x)$.

A simple example illustrating this process is to find N where $F(x) = x^2 - 2$.

$$N(x) = x - \frac{x^2 - 2}{2x} = \frac{x^2 + 2}{2x}.$$

Some time was then spent on when and why Newton's Method fails to find roots: when F' does not exist at the root, when $F'(x) = 0$, when the initial guess is not sufficiently close, when there is no convergence because the orbit is periodic, and when the orbit is chaotic. Examples of each case were presented.

By using the graphing calculator and some prepared examples, we led the students to make the conjecture that the roots of F are the attracting fixed points of N. This was then rigorously established following Devaney's proof of Newton's Fixed Point Theorem.

Finally, we began looking at the complex case by considering $F(z) = z^2 + 1$ where $z = a + bi$. z_0s which are real, pure imaginary, and then complex are used as initial values. Homework assignment sheets with $F(z) = z^2 + 1$ and

$F(z) = z^3 - 1$ were discussed. Computer images of their respective basins of attraction are examined.

This unit lasted for approximately two weeks (8 classes) during which students had reading assignments from both the text and the supplementary sources mentioned above, written assignments using exercises from the text, and assignments using worksheets written by the instructors. Here is a sampling of some of the homework problems.

Problem A. Use the Babylonian square root algorithm to find the square root of 5. Develop a similar root-finding algorithm for the cube root of k. Test your algorithm for $k = 8$, 100, -6. Make a conjecture on a root-finding algorithm for the nth root of k.

Problem B. Find all real roots for each function below. Write the associated Newton Iteration Function, $N(x)$, for each. Find the fixed points of N.

$$F(x) = 4 - 2x$$
$$F(x) = x^2 - 2x$$
$$F(x) = x^4 + x^2$$
$$F(x) = xe^x$$

Note this problem was assigned *before* Newton's Fixed Point Theorem was introduced.

Problem C. Consider $F(z) = z^2 + 1$. Write the associated Newton Iteration Function in three different forms. If $z_0 = a_0 + b_0 i$, find $z_1 = N(z_0)$ in terms of a_0 and b_0. In general, if $z_n = a_n + b_n i$, express z_{n+1} in terms of a_n and b_n. Find z_1, z_2, z_3 for $z_0 = 2 - 2i$. Graph the orbit in the complex plane.

At the end of the unit we assessed the students' understanding of the material with a full period test, consisting of five short answer questions (sample: What is meant by the term "basin of attraction?") and five more substantial problems, including this:

Let $F(x) = x^3 + 2x$, where x is real.

   a. Find N, the associated Newton Iteration Function.

   b. What are the fixed points of N?

   c. Find $N'(x)$.

   d. Show that your answer in part b is attracting.

Note: you must show the substitution and evaluation numerically and not rely on the Newton Fixed Point Theorem.

8   Student projects

8.1   1993–1994 course projects

The first three projects on the Mandelbrot Set, Julia Sets, and Fractals were submitted at the end of the academic year 1993–1994. During that year we focused mainly on real dynamics, including symbolic dynamics. The three projects were assigned by the instructor and the reference materials were: Chapters 16 and 17 from Devaney (1992) sections from Chapter 2 of Peitgen, Jurgens & Saupe (1992b,c), Devaney (1990), Lauwerier (1987) and Mandelbrot (1982).

8.1.1   Julia sets (Doug Sproule)

Doug's project focused on the definition and various properties of Julia sets and was accompanied by images of the Julia sets for $Q_c(z) = z^2 - 1$ and $Q(z) = z^2 - (.12 + .72i)$ generated using programs in BASIC.

Doug began his project with

> It is quite remarkable that Gaston Julia was able to make such advances in the area of complex iterations, considering that there were no computers to enable him to see the results of his work... It is regarded as one of the greatest tragedies of mathematics that the chief creator of these wonderful images died within months of the first computer representations of such sets.

He continued to define Julia and filled Julia sets and dealt mainly with the dynamics of the quadratic family $Q_c(z) = z^2 + c$, for different values of the parameter c.

Next, Doug discussed the quadratic function $Q_0(z) = z^2$ for which the Julia set is the unit circle and the filled Julia set is the unit disk. Referring to some previous results, namely that the doubling function is chaotic, he justified why the squaring function is chaotic on the unit circle. In addition, he explained why after a few iterations any wedge around a point on the unit circle will become larger and larger until it fills the whole plane save the origin. Also he discussed the dynamics of the function $z^2 - 2$ and its Julia set.

Next, Doug introduced the Escape Criterion and explained why the Mandelbrot set is inside a circle of radius 2, centered at the origin. The following corollaries are immediate consequences of the Escape Criterion.

Corollary 1 Suppose that $|c| > 2$. Then the orbit of 0 under $Q_c(z)$ escapes to infinity.

Corollary 2 Suppose $|z| > Max\{|c|, 2\}$. Then $Q_c^n(z) \to \infty$, as $n \to \infty$.

The latter corollary is used to explain why the Julia set of $Q_c(z)$, $(c > 2)$ consists of a Cantor set of points.

Finally Doug focused on the Julia sets of $Q_c(z)$, for $|c| < 2$ and explained how to produce pictures of such sets.

> To create the image, we set a certain number of orbits as the maximum. If after this number of iterations, the orbit is still within the circle [of radius 2], we say that the point is in the filled Julia set. This is an approximation, as are all computer generated images... The color plates that make

the Julia sets such lasting objects in our memories are the result of an analysis of the points whose orbits escape. Each point whose orbit escapes is assigned a color depending on whether it escapes quickly or slowly.... This representation allows us to note several things about a Julia set. While it is by no means a fractal, it contains a self-similarity reminiscent of fractals. If you examine a leaf of a Julia set and magnify it, this leaf will closely resemble the original Julia set.... The massive and sudden loss of regions to the filled Julia sets seems to occur at the bifurcation points ... the understanding of which requires the understanding of another topic in mathematics–the Mandelbrot Set.

8.1.2   The Mandelbrot set (Thomas Holland)

Tom began his project with

> In 1975, the mathematician Benoit Mandelbrot observed that geometric objects that had previously been thought of as pathological aberrations were actually quite ordered and even at times more natural than the stiff geometric shapes of Euclidean Geometry ... In 1980 he discovered the Mandelbrot Set, one of the most interesting and complex shapes in dynamics and indeed in all of mathematics. It is this set which I sought to understand through the use of Robert Devaney's *A First Course in Chaotic Dynamical Systems* ... In order to understand the Mandelbrot set, one must have a thorough understanding of complex functions and their dynamics.

Tom's project focused on the definition and some basic properties of the Mandelbrot set $\mathcal{M}$ for the quadratic function $z^2 + c$, and accompanied the demonstration of key ideas by graphic images.

Tom began with a discussion of the Fundamental Dichotomy Property for Julia sets, namely that Julia sets are either connected or totally disconnected, which in turn can be linked to the fate of the orbit of the point 0. For example, in the case $c = 0$ the Julia set is simply a circle, while in the case $c = -0.75 + 0.1i$ it shatters into a Cantor Set consisting of infinitely many pieces. The observation here is that when the orbit of the critical point 0 remains bounded, the corresponding Julia set is connected, while if the orbit goes to infinity the Julia set is disconnected.

Tom used the Fundamental Dichotomy theorem to introduce the definition of the Mandelbrot set: $\mathcal{M}$ consists of all c values for which the filled Julia set is connected; that is, $\mathcal{M} = \{c \in C : |Q_c^n(0)| \nrightarrow \infty \text{ as } n \to \infty\}$.

Next, Tom provided a derivation of the boundaries of the period one cardioid and the period two bulb of $\mathcal{M}$. This involved the definition of fixed and period two points and the fact that a fixed point z is attracting if $|F'(z)| < 1$, and a point z of a 2-cycle is attracting if $|(F^2)'(z)| < 1$. Thus the boundary of the period one bulb is the cardioid $c = \xi(\theta) = \frac{1}{2}e^{i\theta} - \frac{1}{4}2^{2i\theta}$, while the period two bulb is bounded by the circle $|c + 1| = \frac{1}{4}$. In addition, Tom compared the orbit diagram of the logistic map $L(x) = x^2 + c$ and $\mathcal{M}$.

Tom concluded with

> Hopefully I have captured some of the truly magnificent dynamics and images associated with this beautiful and complicated picture. The relatively new science of Fractal Geometry offers amazing opportunities for the discovery and study of this and other such mathematical wonders. Personally this year's math course has opened my eyes to an exciting new area of math and an entirely different way of approaching the study of math. I've had to think harder during this course than ever before, so I don't plan to stop in this line of work soon.

8.1.3 Fractals (Div Bolar)

Div's paper began with an acknowledgment that there is probably no single definition of the term "fractal" that covers all forms. Some fractals are described as "endlessly repeating geometrical figures," others with the property of self-similarity, others as sets whose fractal dimension exceeds its topological dimension, and still others that "have finite area bounded by infinite perimeters."

After discussing topological dimension and how it can be determined using the notion of hollow spheres and number of points of intersection with the object, Div applied the concept to a point, a line, a square, and a cube, to demonstrate why we assign a dimension of 0, 1, 2, and 3, respectively, to these figures. Div continued by discussing and illustrating other ways of determining fractal dimensions, which are almost always not integers.

The *self-similarity dimension*, introduced earlier in the course, was reiterated as

$$D_s = \frac{\log(\text{number of congruent pieces})}{\log(\text{magnification factor})}.$$

The self-similarity dimension of the Sierpinski triangle illustrated this concept.

Then Div examined the *compass dimension*, using the Koch curve to illustrate this method. For different compass settings y, Div plotted the points

$$(\log(1/y), \log(\text{perimeter})).$$

These points were collinear with slope d. The compass dimension, D_c, is given by $D_c = 1 + d$.

Next, Div defined the *box-counting dimension* and used it to determine the dimension of the Coastline of Great Britain.

Then Div examined the Devil's Staircase, a fractal associated with the Cantor Middle-Thirds Set. "The Devil's Staircase can be visualized as a black image with an infinite number of light beams running through it." This is an interesting example of a fractal because, unlike the earlier examples, its dimension is an integer.

Div concluded with a description of some fractals found in nature—broccoli Romanesco, ferns, the mammalian kidney. He ended by musing

> Perhaps in the near future, and even now, fractals can be used to predict natural occurrences, and to further explore and decipher the ways of the world.

8.2 1994–1994 course projects

The next four projects were written in the 1994–1995 academic year. The first two were assigned to the students with more specific directions. The last two were chosen by the students themselves and worked on almost independently.

8.2.1 The computation of the Feigenbaum constant (Nwang Chollacoop)

Nwang began

> Observing the orbit diagram of the logistic function, Feigenbaum discovered that the ratio of consecutive period doubling intervals converges to a constant. It is approximately $4.6692016091029\ldots$. The question is how to compute the limit of such a messy convergent sequence, because it is not a typical sequence that has an explicit formula for its computation. Furthermore the length of the intervals are very small numbers as the sequence goes to infinity. However, in October 1975, Feigenbaum succeeded in computing the limit. More amazing is that this number is not attached to a specific model. It is universal because it is the same for a wide range of functions.

Nwang discussed a method described on p. 224–227 of Peitgen, Jurgens & Saupe (1992c). Here the computation of the Feigenbaum constant is based on the use of Newton's method. Through this method, instead of bifurcation parameters one approximates the super attractive parameters s_n ($n = 1, 2, 3\ldots$), solutions to the equation $f_a^{2^{n-1}}(p) = p$, where p is the critical point of the function being iterated.

The ratios of two consecutive intervals determined by these parameters approaches a limit—the Feigenbaum constant. Nwang focused on the computation of the first few super attractive parameters for the logistic and the quadratic functions, $f(x) = \lambda x(1 - x)$ with $p = \frac{1}{2}$, and $f(x) = x^2 + c$, with $p = 0$. The process reduces to the use of a system of two recursive equations, applied every time one computes a new approximation of s_n.

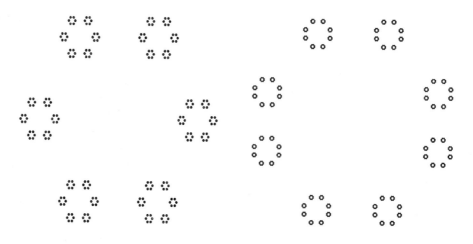

Figure 2: Fractal hexagon and octagon.

8.2.2 Fractals and the TI-82 (Griff Baker)

Griff's project dealt with the creation of fractal objects on the *TI-82* graphing calculator. He started with a discussion of linear transformations of the form (Anton & Rorres (1991)),

$$T_i \begin{bmatrix} x \\ y \end{bmatrix} = s \begin{bmatrix} a_i & b_i \\ c_i & d_i \end{bmatrix} \begin{bmatrix} x \\ y \end{bmatrix} + \begin{bmatrix} e_i \\ f_i \end{bmatrix},$$

where $a_i, b_i, c_i, d_i, e_i, f_i, s$ are scalars and $i = 1, 2, \ldots n$. Here s determines the size of the transformation, $\begin{bmatrix} a_i & b_i \\ c_i & d_i \end{bmatrix}$ determines the rotation … and $\begin{bmatrix} e_i \\ f_i \end{bmatrix}$ determines the translation.

Griff's goal was to create a fractal that is generated by the vertices of a regular *n*-gon. A linear transformation T_i corresponding to each vertex is given by

$$T_i \begin{bmatrix} x \\ y \end{bmatrix} = \frac{1}{n-1} \begin{bmatrix} 1 & 0 \\ 0 & 1 \end{bmatrix} \begin{bmatrix} x \\ y \end{bmatrix} + \begin{bmatrix} e_i \\ f_i \end{bmatrix},$$

where $\begin{bmatrix} e_i \\ f_i \end{bmatrix}$, $i = 1, \ldots, n$, are the vertices of the *n*-gon.

After describing the program used to create the Sierpinski triangle (p. 14–18 of the TI–82 Graphics Calculator Guidebook), Griff modified this program to create regular *n*-gon fractals. For example, the linear transformations for a fractal hexagon are

$$T_i \begin{bmatrix} x \\ y \end{bmatrix} = \frac{1}{5} \begin{bmatrix} 1 & 0 \\ 0 & 1 \end{bmatrix} \begin{bmatrix} x \\ y \end{bmatrix} + \begin{bmatrix} e_i \\ f_i \end{bmatrix},$$

where

$$\begin{bmatrix} e_i \\ f_i \end{bmatrix} = \begin{bmatrix} 0 \\ 0 \end{bmatrix}, \begin{bmatrix} -1/2 \\ \sqrt{3}/2 \end{bmatrix}, \begin{bmatrix} 0 \\ \sqrt{3} \end{bmatrix}, \begin{bmatrix} 1 \\ \sqrt{3} \end{bmatrix}, \begin{bmatrix} 3/2 \\ \sqrt{3}/2 \end{bmatrix}, \begin{bmatrix} 1 \\ 0 \end{bmatrix}.$$

For a fractal octagon they are

$$T_i \begin{bmatrix} x \\ y \end{bmatrix} = \frac{1}{7} \begin{bmatrix} 1 & 0 \\ 0 & 1 \end{bmatrix} \begin{bmatrix} x \\ y \end{bmatrix} + \begin{bmatrix} e_i \\ f_i \end{bmatrix},$$

where

$$\begin{bmatrix} e_i \\ f_i \end{bmatrix} = \begin{bmatrix} 0 \\ 0 \end{bmatrix}, \begin{bmatrix} 1 + \sqrt{2}/2 \\ \sqrt{2}/2 \end{bmatrix}, \begin{bmatrix} 1 + \sqrt{2}/2 \\ \sqrt{2} \end{bmatrix}, \begin{bmatrix} 1 \\ 3\sqrt{2}/2 \end{bmatrix},$$
$$\begin{bmatrix} 0 \\ 3\sqrt{2}/2 \end{bmatrix}, \begin{bmatrix} -\sqrt{2}/2 \\ \sqrt{2} \end{bmatrix}, \begin{bmatrix} -\sqrt{2}/2 \\ \sqrt{2}/2 \end{bmatrix}.$$

Figure 2 shows images obtained by these transformations.

8.2.3 Symmetry and interpretation: A study in fractals (Roberto Lurena and Seth Proctor)

Roberto and Seth began their project with

> The purpose [of this paper] was to study in depth the math and the symmetry behind various fractal images such as the Mandelbrot set(s), Julia Sets, and complex graphs for Newton's method. We built a program that allowed us to draw these fractal images and to further study symmetry and variations based on the power of polynomial functions and other uncommon equations.

First Roberto and Seth took up the subject of the Mandelbrot set and higher order Mandelbrot sets. They described the process of obtaining a computer image for the Mandelbrot set of the quadratic function, then focused on applying that same process to generate n^{th} degree Mandelbrot sets for the functions $z^n + c$. Computer generated images of the second, third, fourth and fifth degree Mandelbrot sets were compared and some common features noted.

> What we discovered primarily was that the number of large bulbs (the regular Mandelbrot set has one large bulb) is $n - 1$. That is to say that the fourth degree Mandelbrot set has three major bulbs, and the fifth degree Mandelbrot set has four major bulbs.

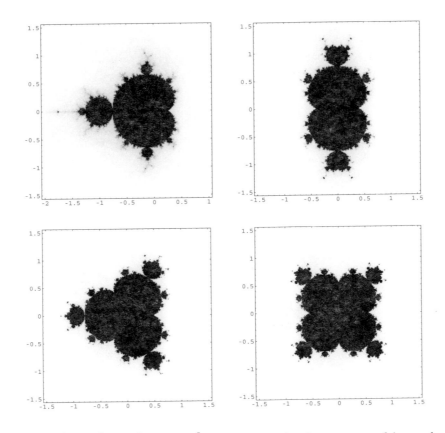

Figure 3: Mandelbrot sets for $z^2 + c$, $z^3 + c$, $z^4 + c$, and $z^5 + c$, demonstrating the symmetry of the number of pieces in relation to the degree of the polynomial.

They also observed that

> all Mandelbrot sets are symmetric about the real axis, that odd n degree images are also symmetric about the imaginary axis, and that there exists a polar kind of symmetry for all Mandelbrot sets by the amount $360/(n - 1)$ degrees.

In addition, they discussed an attempt to distinguish the bulbs of the Mandelbrot set by assigning different colors to different period cycles to which the orbit of the critical point converged. If no periodicity was observed, the parameter value was colored white. They arrived at an interesting observation: the computer draws a perfect circle in the center of the main bulb. Increasing the accuracy with which the periodicity was determined made the circle grow larger. Also, "The farther a point is from $(0, 0)$, the more iterations it takes for the orbit to converge to a period 1."

In the section on the Julia sets, Roberto and Seth described some observations common for the Julia sets of $z^2 + c$, $z^3 + c$, $z^4 + c$, $z^5 + c$, $cz(1 - z)$, $c \sin z$, $A \exp z$.

In the third section on Newton's method, they focused on how to create computer images of the basins of attraction. Their program dealt with Newton's method for the functions $z^2 + 1$, $z^2 - 1$, $z^3 + 1$, $z^3 - 1$, $z^4 + 1$, $z^4 - 1$. A simplification

of Newton's method for $z^n + c$ is

$$N(z) = \frac{1}{n} \left[(n - 1)z - \frac{c(\bar{z})^{n-1}}{(a^2 + b^2)^{n-1}} \right]$$

where $z = a + ib$.

In the fourth section, Roberto and Seth described a fractal movie they made.

> Having looked at various fractals... we decided to try our hand at making a small time movie... What we chose to film was something similar to what we saw in the Devaney films this year, when various parameter values were taken outside the Mandelbrot set and through the axis. These parameters were then used to draw the corresponding Julia sets. We took the same approach, only we used the function $y = z^3 + c$ instead of $y = z^2 + c$. What we hoped to observe was the same inward explosion at particular values on the graph, and the same spiraling and filling/emptying of color from the Julia set. Needless to say we were not disappointed.

Needless to say their teachers were not disappointed either.

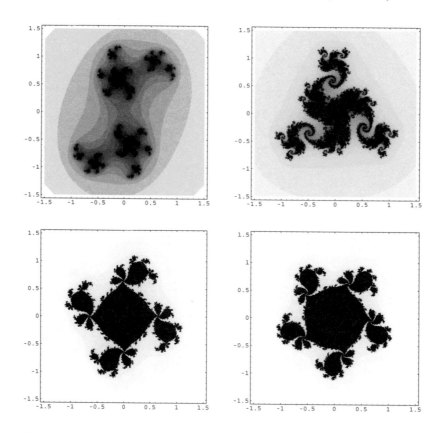

Figure 4: Julia sets for $z^2 + c$, $z^3 + c$, $z^4 + c$, and $z^5 + c$, for different c. They demonstrate a clear symmetry, and number of "pieces," related to the degree of the polynomial.

8.2.4 Newton's function for complex functions and the TI-82 graphing calculator (Sarah Craig and Sarah McDougal)

Another pair of students elected to do their project on Newton's Method, attempting to write programs for the *TI-82* that would produce the orbit of any seed for the Newton Iteration Function associated with $Q_c(z) = z^n + c$, $S(z) = \sin^n(z)$, and $C(z) = \cos^n(z)$, where z is complex, c is real, and n is a positive integer. Sarah and Sarah used the simplification of Newton's method described in section 8.2.3. About it, they said

> This is a fairly easy expression to evaluate, with the exception of z^{n-1}. This is because as n gets larger, the coefficients of the expansion change. To fix this problem [one of the authors] suggested that we use brute force in calculating the real and imaginary parts of z^{n-1}. We therefore wrote a supplemental program to run inside the main program that simply multiplies $(a + bi)$ times itself $n - 1$ times.

The complex sine and cosine functions presented a more interesting challenge. Sarah and Sarah computed the Newton Iteration Function associated with $\sin^n(z)$ and $\cos^n(z)$ to

be $N(z) = z - \tan(z)/n$ and $N(z) = z + \cot(z)/n$, respectively. Iterating the tangent and cotangent functions, however, proved not to be so easy. To determine simplified iteration formulas for the real and imaginary parts of these functions, they had to perform substantial calculations with pencil and paper.

All of this work is then translated into three programs on the *TI-82*, one for each of the Newton Iteration Functions associated with $Q_c(z) = z^n + c$, $S(z) = \sin^n(z)$, and $C(z) = \cos^n(z)$.

9   What students say

In order to better evaluate the progress of our program in fractals and dynamical systems, we asked some of our students—after they had graduated—about their experience. We used these questions as guidelines.

1. What aspects of your experience in the *Dynamical Systems and Fractals* course did you find most exciting and why?

2. What was your project about and which part of it did you enjoy most?

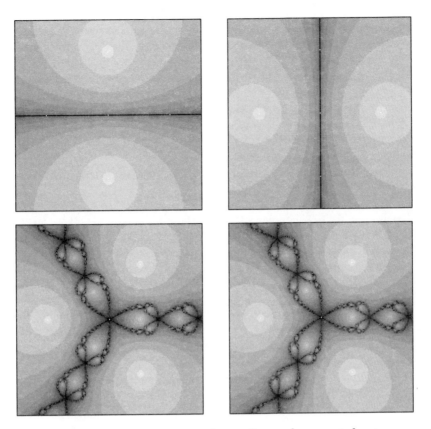

Figure 5: Newton fractals for $z^2 + 1$, $z^2 - 1$, $z^3 + 1$, and $z^3 - 1$.

3. Are you interested in continuing the study of Fractals and Dynamical Systems at your university? If so, did you find an adequate course offering? If there is no such course offering do you think there should be one?

4. Do you think that this kind of course could open further gates to understanding different phenomena in nature? in life? in sciences? Why?

5. Any suggestions concerning the course?

Here is a sampling of responses.

"The hardest part of the course was not the calculations, but understanding the concepts. Yet at the same time it was the concepts that made the course so intriguing and tantalizing. Just from an artistic point of view fractals are beautiful, but when you start to realize their connections with mathematics, and with nature, it is truly astounding."

"Before I took this class, I didn't know anything about chaos, but now I do. I liked when we were watching videos relating to fractals or the computer programs that allowed us to visualize what we learned in the textbook."

"Although I came into the course with a strong mathematical background, I was immersed in a world of math completely new and interesting to me. I was equally impressed by how the course focused on the theory of chaos, as well as the rigorous mathematics behind it."

"There is something indescribably exciting about learning something in the class, and then going on to another topic only to find the same figures popping up again. Basically, there are few things that I have ever studied that are as exciting or uniquely intriguing as the study of *Dynamical Systems and Fractals.* "

"The aspects that I enjoyed most about the *Fractals and Dynamical Systems* course were the material itself and the class environment. The material was exciting both from a mathematical standpoint and because it seemed new and relevant. Most people dismiss higher math as being something totally irrelevant to the real world. With fractals, however, people see that it can shed light on nature, on how weather, with all our technology, can still never be predicted more than a couple of days in advance. The material was as challenging as any I had ever done, but it was certainly rewarding to be a part of the budding math of fractals and chaotic systems."

"For my project, another student and I wrote a Macintosh application that graphed full fractals (Mandelbrot sets, Julia sets, Newton's Method). This required calculating many equations, and really gave me a good understanding of what was happening behind the pictures, as well as the mathematical complexity of the images. It gave us a chance to really explore, and just play around with both images and the math to see what we could find. Really, I just enjoyed being able to head into a project to see what I'd get out of it."

"In my final project I investigated fractal geometry and the uncanny correlations between fractals and objects in nature.... Finding concrete examples for my research paper was probably the most exciting and enjoyable aspect of the project."

"My project, on the quadratic function in its complex form and the Mandelbrot set, was the culmination of my work in the course. I was able to work through the math behind a beautiful fractal picture and present it to a larger audience. I think the fact that we were able to share our projects with the rest of the math department plus some other interested faculty and students was one of the best aspects of the projects."

"The materials, the trip to Boston [for Bob Devaney's Dynamical Systems Field Day at B.U.], the final project, and the lecture with Bob Devaney [were highlights.]"

"As for explaining nature, I think that we already understand more after the work of people like Lorenz, and I expect that given how many branches of study are exploring chaotic systems, we shall see more answers emerging in the future based on these studies."

"The course at Hotchkiss was one which was a very positive experience for me. There will continue to be new developments with ramifications in all sorts of disciplines, and it is a wonderful thing for a student to feel like he or she can be involved in and understand such new material in the world of math. I remember my own amazement when I was told that Mandelbrot was alive and well and even could be contacted for discussion about math. It had never occurred to me that I could be a part of something in math that was that current.

"The course was by far the most rewarding course that I took at Hotchkiss."

10   Conclusion

Receiving this kind of feedback (pardon the pun) is both promising and satisfying. Our short experience with this course suggests that some mathematics students yearn for such an experience. What we wanted to do, and think we have done, is to whet some appetites for higher level contemporary mathematics. If nothing else, our courses have shown to students that mathematics did not stop with the development of calculus in the 17th century, but, like Professor Mandelbrot, is "alive and well," even today. We think that this material is accessible—at the appropriate level—to many high school students, as well as freshman and sophomore college students, and that they will find its study exciting, rewarding, and relevant.

Chapter 6

Reflecting on Wada Basins:
Some Fractals with a Twist

Dane Camp

As a secondary mathematics teacher for over twenty years, I can attest to students' fascination with fractal geometry. Perhaps they simply mirror my enthusiasm for a subject whose wonderful connections are continuously unfolding. However, I contend the reasons are much deeper, for the ubiquity of fractals across various disciplines has led me to keep an ever expanding scrapbook of fractal applications. In fact, the occurrences of fractals in our culture are so pervasive that, when I give my students an iteration project every year, they always uncover new and surprising manifestations. Such an experience happened to me when one of my colleagues in the physics department at our high school asked me about *Wada Basins*, a special class of fractals, unfamiliar to me, that he had stumbled across on the internet.

Upon investigation, I found that Wada Basins are, informally speaking, attractors whose fractal boundaries border all attractors simultaneously. Though described first in 1917 by the Japanese mathematician Yoneyama, who devised them as particular topological constructions (see pgs 143–145 of Hocking & Young (1961)), the *Lakes of Wada* now can provide a connection between magnetic fields, Newton's method for finding roots, and the chaotic scattering of light. Mathematical properties of Wada Basins are explored in Kennedy & Yorke (1991) and in Nusse & Yorke (1996).

What follows is a lesson used in my AB Calculus class to reveal these connections. Any class that is exposed to a basic knowledge of derivatives and has a grasp of operations with complex numbers should be able to complete the investigation successfully. The lesson is divided into four parts:

   I. A poetic definition of Wada basins, since a formal definition is not required.

   II. A short description of the attractive pendulum.

   III. An investigation of computing the cube roots of unity using Newton's method.

   IV. A demonstration of the chaotic scattering of light in mirrored spheres.

The entire investigation took slightly less than two 40 minute class periods. The first period ended as students were working in groups on the activity sheet. They completed the sheet as homework. The following day, the results were compiled and the remainder of the lesson was completed.

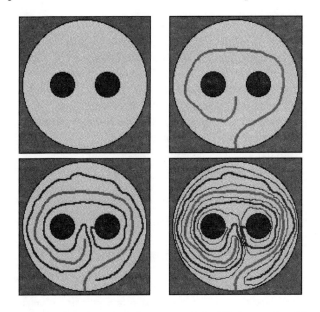

Figure 1: The first three branches of the Wada canals. After infinitely many have been dug, every point of the island is on the edge of a canal to the ocean and to each lake. See the color plates.

1   The Lakes of Wada

Once upon a time, long ago, yada yada yada,
three siblings lived on an island called Wada.
Julia, Hilbert, and Cantor all got along,
and everything was perfect, for nothing went wrong.
This island was but a few moments wide,
with two little lakes nestled inside.
The surrounding oceans where Cantor set out to play
singing old sea shanties and sailing all day.
Julia loved the red lake where she'd dangle her feet,
for it was just like her, it was gentle and sweet.
Hilbert's space was the green lake for he loved to fish.
Casting about from the shoreline as long as he'd wish.
However they all craved each other's presence
'cause a few moments away can be quite a distance.
One day Cantor said, "Hey sister, hey brother my pal
would you both mind if I'd dig a canal?
That way the ocean would always be near
and I'd still be close to the ones I hold dear!"
Of course, they agreed for his motives were true
but asked, "When you've finished, can we dig one too?"
So Cantor cut a channel, which took him all day,
till no point on Wada was over half a moment away.
The next day sweet Julia set out with a spade
and a channel 1/3 moment from anywhere she made.
Hilbert, as agreed, then dug his own ditch
so that a quarter moment from anywhere his line he could pitch.
It all worked so well as each channel went through,
so they all decided, why not continue?
Cantor extended his to within one fifth,
then Julia and Hilbert etcetera and forthwith.
So each day they took turns extending the trend
and so on and so forth forever without end.
Yet when all was finished Julia said, "Look now brothers,
each boundary between waters borders on the others!"
The edge twixt any two waters, though it sounds absurd,
always, upon inspection, doth border the THIRD!
So from that moment on they were in constant touch,
for every boundary between two waters the other would clutch.
So tell all your friends from Seville to Granada
that's the true story of the Lakes of Wada.

2   The attractive pendulum

Consider a pendulum with a metallic bob. Below the pendulum place three colored magnets, one red, one blue, and one yellow. When the bob is released, it will naturally be attracted to the magnets, but its momentum also will affect the trajectory. Though the bob will swing around for a while, eventually it will hover above one of the magnets and become

trapped. Now suppose you colored the point immediately below the position where you released the bob the same color as the magnet over which it eventually stopped. For example, if the bob was trapped above the red magnet, you would color red the point below where you released the bob. If you were to continue to do this for every point within the vicinity below the bob, how would the pattern look? Imagine it and describe what you see. Why do you suppose it would look that way?

The actual experiment would be very time consuming, and inaccurate, to do by hand. Nevertheless, a few tries are quite entertaining. "Romp," a pendulum configured in this way, can be purchased at

`http://www.hogwildtoys.com`

To see the whole picture, the experiment can be simulated by a computer program, for example, the Windows program "Chaos," available at

`http://www.mathcs.sjsu.edu/faculty/`
`rucker/chaos.htm`

The result, shown below, might surprise you. Perhaps you imagined that the area would be divided up into three zones, yet the boundaries between these zones seem rather strange. In fact, it can be demonstrated that any point on a boundary actually touches all three colors simultaneously! If you were to zoom in on the boundary, you would see more and more detail, each picture including all three colors. See pages 758–765 of Peitgen, Jurgens, & Saupe (1992a).

Figure 2: The basins of attraction of an iron pendulum suspended over three magnets, in each of the largest regions. Starting from all points in the yellow region, the pendulum comes to rest over the yellow magnet, and similarly for the other two colors. Every point on the boundary between two regions is also on the boundary of the third.

3 Newton's method for finding cube roots of unity

The exploration sheets in Appendix 1 allow students to discover Newton's method for finding the cube roots of unity by using iteration. I found it easiest to divide the students into groups and allow them to work together. Though the materials are intended to stand alone, it is essential that you circulate among the groups to assist them with the finer points of the lesson. The following day, after you have compiled the results and discussed what is going on in the figure, you should ask where they have seen something similar before. We hope they will remember the computer generated figure for the attractive pendulum! The computer also can provide a stunning picture of the cube root of unity attractors. Again, if you were to zoom in on any boundary point, you would find that it borders all basins of attraction simultaneously.

Figure 3: The basins of attraction of Newton's method for finding the roots of $z^3 - 1$. Starting from all points in the yellow region, Newton's method converges to the root in the yellow region, and similarly for the other two colors. Every point on the boundary between two regions is also on the boundary of the third.

Computer pictures can be generated with Windows program "Fractint," available at

http://spanky.triumf.ca/fractint.fractint.htm

For background on Newton's method consult pages 223–238 of Goodman (1974). Wada basins for Newton's method are described on pages 771–775 of Peitgen, Jurgens, & Saupe (1992a).

4 Chaotic scattering of light by mirrored spheres

Take four mirrored spheres and place them in a tetrahedral configuration with three spheres touching on the bottom and the last nestled on top. Though gazing globe lawn ornaments are ideal, small Christmas ornaments will suffice. Place the necks of the three bottom spheres into something stable, such as cups, before setting up the configuration. Make sure the neck of the top sphere stays away from the center of the configuration.

Ambient light should be enough to convince you that something special is happening in the interior between the four spheres. However, imagine that you shine three different colored lights into three of the four openings, and then look through the other opening. What do you think you would see? Describe it.

If you shine red light into one opening, blue light into another, and white light into the third, from the remaining opening a dazzling figure will appear. Though the figure looks strangely similar to the Sierpinski gasket, it is yet another Wada Basin. A simple investigation demonstrates this: Take a standard laser pointer and shine it through the observation opening on a patch of blue light, for instance. The laser beam will exit from the side into which the blue light shines. The same will hold true for any patch upon which you shine the laser light. However, what do you suppose happens when you shine the laser on the boundary between two patches? The laser light will exit through all three openings simultaneously.

On the right of Figure 4 is the pattern illuminated by just the white light through one opening between the spheres. On the left of Figure 5 red and blue light is shined through two of the other openings. On the right of Figure 5 is the pattern produced by all three lights. The pictures in this section came from David Sweet's website

http://www.chaos.umd.edu/Spheres_Photos/
spheres_photo.html

See also Camp, Chiaverina, & Senior (1999) and Sweet, Ott, & Yorke (1999).

5 Making connections

This investigation into the special class of fractals called Wada Basins illustrates both the wonder and power of mathematics. The fact that connections can be found between the seemingly dissimilar areas of magnetism, function iteration, and reflecting light reveals that mathematics is a field that will always contain wonder for those willing to explore. Because these are practical applications that have a phys-

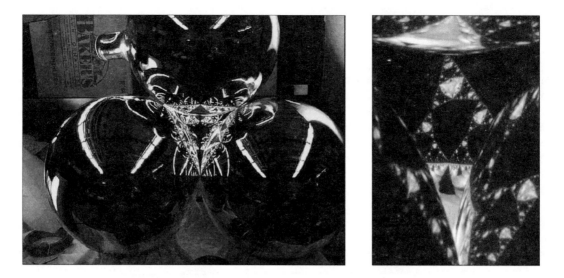

Figure 4: Left: the set-up for the chaotic light scattering experiment: four mirrored spheres arranged in a tetrahedron. Right: the pattern revealed when white light shines through one opening between the spheres.

Figure 5: Left: the pattern revealed when blue and red light shines through two openings. Right: all three lights illuminating the interior of the region. See the color plates.

ical component, they also demonstrate that making mathematical connections is a powerful tool for understanding the world around us. That such lessons can be learned in the secondary school classroom everyday, should keep the teaching of mathematics an endeavor continually sprinkled with intriguing marvels.

Appendix 1
Newton's method worksheets

Newton's method for finding wada basins

Graph the function $f(x) = x^3 - 1$.

1. What appears to be the x-intercept of the graph?

2. What is the slope of the graph of $f(x)$ at $x_1 = 2$?
3. Determine $f_1(x)$, the line tangent to $f(x)$ at $x_1 = 2$.
4. Sketch $f_1(x)$ and determine its x-intercept. Call it x_2.
5. What is the slope of the graph of $f(x)$ at x_2?
6. Determine $f_2(x)$, the line tangent to $f(x)$ at x_2.
7. Sketch $f_2(x)$ and determine its x-intercept. Call it x_3.
8. If you were to continue this procedure, what do you think would eventually happen to the value of x_n?

Determining a general solution

9. What is the slope of the graph of $f(x)$ at x_n?
10. Determine $f_n(x)$, the line tangent to $f(x)$ at x_n.
11. Determine the x-intercept of $f_n(x)$.

12. Replace x_n in the expression above with the calculator variable ANS. Let $x_1 =$ANS and repeat the expression to complete this chart.

x_1	x_2	x_3	x_4	x_5	x_6	x_7	x_8	x_9	$x_{10}...x_n$
3									
-2									
10									

Wada Basins

1. Type the complex number $-1 + 1i$ into your calculator. (On the TI83 use 2nd. to get i; on the TI86 use $(-1, 1)$.) Push enter and repeat the procedure

$$(2\text{ANS} \wedge 3 + 1)/(3\text{ANS} \wedge 2).$$

   What happens?

2. The function $f(x) = x^3 - 1$ actually has three complex roots. The obvious one was discussed on the first page of the worksheet. You discovered another root in problem 1 above. There is one more, the conjugate of the solution above. (Use $-1 - i$ to see it appear.) Complex numbers that are used as the initial value of ANS can go to *any one* of these three roots. Below is a chart of complex numbers that we will color according to which root they eventually become under iteration. The complex number $-1 + 1i$ is in the upper left-hand corner of the grid. We will color the grid in this fashion. If the initial point eventually goes to 1, we will color the corresponding box red. If it eventually goes to the root you found in problem 1, color it yellow. If it eventually goes to the conjugate of this root, color it blue. Complete the grid.

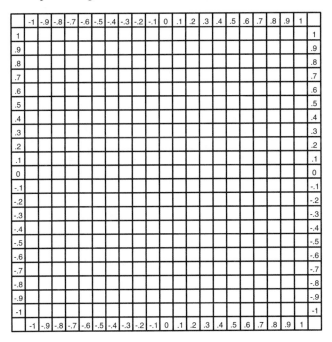

Appendix 2
Newton's method worksheet answers

1. $x = 1$
2. 12
3. $y = 12x - 7$
4. 17/12
5. 289/48
6. $y - 3185/1728 = (289/48)(x - 17/12)$
7. approximately 1.11.
8. It seems to go to the root 1.
9. $3(x_n^2)$
10. $y - (x_n^3 - 1) = 3(x_n^2)(x - x_n)$
11. $x_n - \dfrac{x_n^3 - 1}{3(x_n^2)} = \dfrac{2(x_n^3) + 1}{3(x_n^2)}$
12.

2.04	1.44	1.12	1.01	1.00	1.00	1.00	1.00	1.00	1.00
-1.25	-0.62	0.45	1.92	1.09	1.01	1.00	1.00	1.00	1.00
6.76	4.45	2.99	2.03	1.43	1.12	1.01	1.00	1.00	1.00

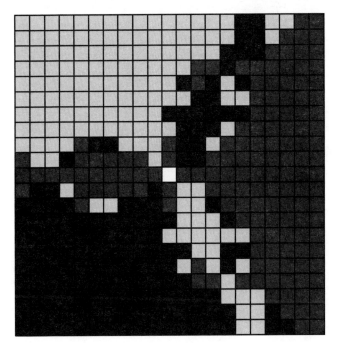

Chapter 7

Learning and Teaching about Fractals

Donald M. Davis

I have been a professor since 1972, and have published many papers on my research in algebraic topology. During the past ten years, teaching has become increasingly important to me, and fractals have had a lot to do with that. In this chapter, I will discuss how I learned about fractals, and how I passed this learning on to a wide variety of people.

1 Early encounters with fractals

My introduction to fractals came in May 1984 at the Cornell Topology Festival, a large annual research conference at which I was one of the invited speakers. During the lunch break, John Hubbard entertained many of us with some of his blowups of the Mandelbrot set. This was the first time that I had heard of this concept. I believe that Hubbard told us the definition of the M set, and perhaps a bit more. I remember being very struck by the strange beauty of these photographs, and within a year my living room wall was adorned with a framed print of part of the M set with lightning shooting out into a red background. I enjoyed explaining to dinner guests a little bit of the mathematics behind this picture, and in particular that the baby bug out in the lightning was really attached to the main body by a thin filament, even though we couldn't see the filament. During this period, 1984–1986, my early years of fractal appreciation, I read the *Scientific American* article by Dewdney, and wrote BASIC programs to create the M set in multihour runs on my PC.

Although Lehigh University, where I have taught since 1974, is generally thought of as an engineering school, there are about 1500 undergraduates who major in humanities or social science, and they have a distribution requirement of at least one semester of mathematics. For a few years, their requirement was two math courses. In the early 1980s, I introduced two courses, *Introduction to Finite Mathematics* and *Introduction to Mathematical Thought*, to service many of these students who preferred not to take calculus. In the latter

course, I could teach whatever topics I thought might catch students' fancy. In Spring 1988, I tried fractals for the first time, devoting four lectures, including one that introduced complex numbers and one in a computer lab. The goal was to get them to understand the definition of the M set and how a computer can magnify a small portion of the set. I was disappointed with the results when an exam showed that most of them didn't understand. But after the course was over, I received a note from an art major saying how much she had enjoyed the course, and that she had learned useful things.

During Summer 1988, I saw Peitgen & Richter (1986) at the home of a mathematician friend, and was taken by the beautiful blowup sequence of the M set and also by the somewhat obscure mathematical discussions. I bought a copy and showed it to some of my colleagues. One of them told Lehigh's Director of Galleries about it, and he asked whether I could obtain such pictures for an exhibit. Obtaining prints from Art Matrix and Media Magic, I became Curator of an exhibit of fractal art in one of Lehigh's galleries during the Spring of 1989. This exhibit obtained extensive publicity, with full color pictures of fractals (and a small black and white of me) occupying most of the front page of the D section of the *Morning Call*, the newspaper that virtually everyone in Allentown and Bethlehem reads. The local television station did a nice interview featuring the exhibit.

Paul Martino, an eighth grade student, was told about the *Morning Call* article by one of his teachers. Paul had a science project on fractals, and called me to see if he could meet with me in the evening to see my exhibit and talk about fractals. We did this, and it seemed as if Paul knew almost as much about fractals as I did. During each of his three high school years, Paul had a math-related science project, and would always come up to Lehigh with his father to tell me about his project and to look for materials in Lehigh's library. They were impressed by all the studying going on in the library and also by this professor who would spend an evening with them, and so Paul ended up applying only to Lehigh,

Harvard, and Princeton. A Trustees' Fellowship covering all of his expenses led him to choose Lehigh, and he was a phenom here. His 3.96 GPA ranked him first in the College of Arts and Sciences, where he was a Computer Science major with a minor in mathematics. He was in several Scholars programs at Lehigh that enabled him to spend a significant amount of time with people such as Lee Iacocca and Dan Quayle. He graduated from Lehigh in three years in Spring 1995, and attended one of the top computer science graduate programs in the country. His life would have been different had it not been for the article about my fractal exhibit.

2   Later experiences

2.1   Liberal arts teaching

Spring of 1989 was when fractals started taking over a large part of my life. In addition to the exhibit, I gave two courses and several demonstrations for the general public. One of the courses was the same *Intro to Math Thought* course that had included three lectures on the M set in 1988. This time I devoted six lectures to fractals, which seemed to help the students' understanding of the M set. By this time, I was starting to think about writing a book for my course. Each year, I would try a different text, but would always add so much of my own material that the students would complain "If you weren't going to use the book, why did you make us buy it?" I started writing detailed lecture notes and using them instead of a text. The three main topics that I had found most effective were non-Euclidean geometry, Number Theory and Cryptography, and Fractals. In 1989, I tried covering all of them, but was still disappointed with the amount that the students learned. In subsequent years I would cover just one or two of the topics. My lecture notes kept growing and were finally published in 1993 by Princeton University Press as *The Nature and Power of Mathematics* (Davis (1993)). For several years I have used this book for the course, though recently I have diverged somewhat from the text, primarily to use Fractint to replace BASIC programs.

My book contains five chapters.

1. Some Greek Mathematics
   - 1.1 π and Irrational Numbers
   - 1.2 Euclidean Geometry
   - 1.3 Greek Mathematics and Kepler
2. Non-Euclidean Geometry
   - 2.1 Formal Axiom Systems
   - 2.2 Precursors to Non-Euclidean Geometry
   - 2.3 Hyperbolic Geometry
   - 2.4 Spherical Geometry
   - 2.5 Models of Hyperbolic Geometry
   - 2.6 The Geometry of the Universe

3. Number Theory
   - 3.1 Prime Numbers
   - 3.2 The Euclidean Algorithm
   - 3.3 Congruence Arithmetic
   - 3.4 The Little Fermat Theorem
4. Cryptography
   - 4.1 Some Basic Methods of Cryptography
   - 4.2 Public Key Cryptography
5. Fractals
   - 5.1 Fractal Dimension
   - 5.2 Iteration and Computers
   - 5.3 Mandelbrot and Julia Sets

Written for good liberal arts students, it could also work for a course for good high school students, or for college freshmen thinking about a possible math major, or for the interested non-student. For a one-semester course I recommend one of the following

 i. 1.1, 1.2, and Chapter 2
 ii. 1.1, 1.3, and Chapter 5
 iii. Chapters 3 and 4

If I become frustrated when students don't learn the concepts or complain about the computer assignments, I can take solace from the occasional student who lets me know that he or she appreciates it. In December 1994 I received a letter from a student who had graduated that May, and had been in my *Intro to Math Thought* course in 1992. He explained in his letter that he was driving his mother somewhere, and she asked him what he had gained by passing a certain car. He told her that other cars might come in between them, and that by getting through some traffic lights that the other car misses, he might arrive at his destination many minutes ahead of the other car. He compared this to what I had taught them about the butterfly flapping its wings and causing a change in the weather far away. He then mentioned *Jurassic Park* as another application of chaos theory, and closed by writing "Aside from all this, I just wanted to tell you that your Math 5 class was the most interesting class I had at Lehigh." Teachers need an occasional letter like that.

2.2   Fractals seminar

The other fractals course I gave in Spring 1989 was a once-a-week seminar on the mathematics of fractals, attended by about 15 people, including undergraduates, graduate students, and professors from the mathematics, physics, computer science, and engineering departments. During the first half of the semester I focused on Barnsley (1988). I felt that an excellent example of abstract mathematics being used in an interesting way was applying the Contraction Mapping Theorem for metric spaces to guarantee that an iterated function system

(IFS) has a unique attractor. The second half of the semester was devoted to Julia sets and the M set, only scratching the surface of the theoretical treatment of the various criteria for Julia sets. I shared some of my Pascal programs for IFSs and Julia sets with the seminar participants. I think they found IFSs most interesting.

One student in the seminar was my Ph.D. student, Ken Monks[1]. His thesis work was in algebraic topology, but he was also an excellent computer programmer, having worked in industry for several years between undergraduate and graduate school. His C programs for IFSs and Julia sets were much more effective than my Pascal programs, and added much to the class. Ken has maintained an interest in using fractals in teaching during his years at The University of Scranton. We have exchanged many materials for courses on fractals. Ken introduced me to Fractint, a remarkably fast and versatile suite of fractal programs.

Ken also was invaluable in the preparation of my book, Davis (1993). One of his contributions was to think of the name of the book, *The Nature and Power of Mathematics*, after my editor and Benoit Mandelbrot both disapproved of my original title, *Selecta Mathematica: Selected topics in mathematics for liberal arts students*. A more substantial contribution was the job of proofreading the manuscript. Ken's biggest contribution was producing the sequence of blowups of the Mandelbrot set that appear in the book. He did everything: finding an interesting region on which to focus, letting his 386 machine generate extraordinarily detailed pictures on overnight runs, and experimenting with different camera settings while photographing the screen. Despite these efforts, we were both disappointed with the quality of the photos that ultimately appeared in the book. Much of the brilliance of Ken's colors was lost. But still his sequence of blowups has a number of features that I haven't seen anywhere else. He focuses on the cusp of the M set and shows that it is an extremely sharp crevice. Inside it, he shows some highly asymmetrical baby bugs, and then focuses on a region that seems to be far away from the M set, but after lots of blowing up, we finally see a baby bug at a magnification of about 10^{14}, surrounded by a spectacular array of colors.

3 General lectures

Also in the Spring of 1989 I realized how effective the topic of fractals is for general lectures. Associated with the art exhibit, I gave a slide show that attracted about 100 people, mostly from Lehigh's art department. Also, I gave the first of three annual lectures to prospective students and their parents at Lehigh's Candidates' Day. I used a computer projection system to show these people the Koch curve, a fractal fern, and the Mandelbrot set, and I explained some of their properties.

---

[1]To see how near the tree the acorn has landed, see Ken Monks' chapter in this volume. *Eds.*

This was a very popular talk, as I learned from some of these students after they matriculated at Lehigh.

Two years later, I gave a demo and talk to Math Conn, a science symposium for middle school girls at nearby Cedar Crest College. I let the girls try some hands-on manipulation with Fractint, and then gave my slide show to a group of some 200 students. I received a good deal of newspaper publicity for this. I can't imagine another topic in mathematics being so effective for such a presentation, and I would be surprised if that did not nudge some girls toward a career in mathematics.

I also gave a similar talk at my church. During the summer, our Unitarian Church changes its format to lay-led services of a not-necessarily-religious theme. One member of the congregation teaches at a high school about 25 miles away. He told a math teacher there about my talk, and she contacted me to see if I could give a presentation to her gifted class. By that time, I was using Fractint for such presentations, and her class really enjoyed it. She showed her appreciation by giving me a Fractals calendar as a present that year, and Lightman (1994) a year later.

My first meeting with Benoit Mandelbrot ultimately led to a similar relationship with another high school teacher. In Spring 1991, I drove to Princeton to hear Mandelbrot give two talks. Between talks, I gave him a copy of the fractals part of a draft of my book. His hosts Fred Almgren and Jean Taylor, were able to incorporate me into the dinner party they were having for Mandelbrot and some of their students. This gave me an excellent opportunity to talk to the man. Some months later I received a telephone call from a high school teacher about 60 miles away who had been told about me by Mandelbrot. This teacher had decided to introduce his students to fractals and had the gumption to call Mandelbrot to ask how to do it. Mandelbrot told him that I lived near him and had written a book about fractals, and suggested that he contact me. We arranged for this teacher to visit me at Lehigh, and we stayed in contact for some time after that. This teacher, Bob Swaim of Souderton High School, received national publicity, including an appearance on "Good Morning, America," in January, 1995, for pointing out that a Boston Chicken advertisement was confusing permutations and combinations.

3.1 Other fractals courses

I have given two other types of fractals courses at Lehigh, which have disappointed me slightly in my inability to get through to most of the students. One was a course in Fractal Geometry for math majors which I gave in Spring 1994. We studied IFSs and fractal dimension from the main text, Barnsley (1988). I used a small portion of Falconer (1990) for the theory of Julia sets and the M set. I find Barnsley's treatment of this topic a little bit quirky. This was a real math course, with minimal computer illustrations. I expected the students to prove theorems; they were math majors, after all. There were nine students in the course. Two of them were clearly better students than the others, but it wasn't until the final exam that I realized the extent to which this was the case. I thought that I was gearing the course toward the lower echelon, but all except the two good students did very poorly on the final. I realized that most of our math majors are not as adept at proofs as I would like. So, in later versions of the course, I adjusted my sights a little.

The other course is a freshman seminar called *Fractals: unexpected beauty in mathematics*. I gave it in the Spring and Fall semesters of 1993. Lehigh's College of Arts and Sciences requires all freshmen take a freshman seminar, an ordinary 3-credit course, but generally in a nonstandard topic and in a class of no more than 16 students. One purpose of the seminar is to encourage discussion rather than lectures. My seminar failed in this regard. I had it in a room in which every student was seated at a computer, networked together. The computers were used in virtually every class meeting. Fractint was the primary software, although many of the BASIC programs from my book also were used. The problem with the computers was that the students could hide behind them, avoiding any discussions. Seminars in other subjects were held in a roundtable format, so students could hardly avoid talking. In their evaluations, many students complained about the lack of discussion, but I am not sure that they really wanted it. The math professor who taught a freshman seminar prior to me said that the students only seemed happy when he would lecture to them. He is the most popular teacher in the department, and was giving a seminar on *The Art of Mathematics*. If he couldn't engage the students in conversation, the problem may be the nature of the subject.

One problem with the seminars was hardware-related. The students had to create many figures using Fractint, and then print the result of their labors. The network on which Fractint was installed often was overloaded with users, making it hard to print pictures. It seems that Fractint's printing option, at least as we had it installed, was somewhat ornery, even if there were not other users. The third difficulty with the seminars was that, as usual, I overestimated the abilities of the lower half of the class. Even though this was not a high-powered math course, students would still get lost and be afraid to ask for help. Subsequent topics would then be wasted on these students. The course really had four almost-independent topics: Fractal dimension and L-systems, IFSs, chaos, and the M set. On each topic, students could almost start afresh, but within a topic, they could really get lost.

Of course, there were good things about the seminars. I think the best was having the students create their own L-systems and IFSs using Fractint. This gave them a lot of room to be creative, more than in any other math course I have taught. It was fun for them to share their creations with the class. I intend to do this in my *Intro to Math Thought* class, although we won't have individual computers. If I revise my book, I will certainly include more of these creative activities. The IFSs were particularly successful. The first

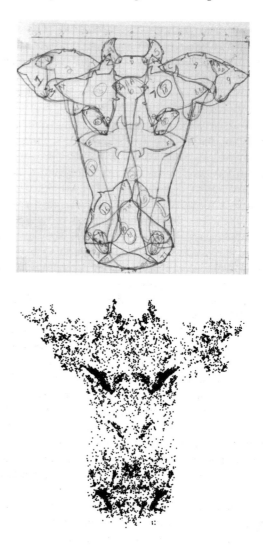

Figure 1: Top: A cow covering. Bottom: A fractal cow

time I gave the course, one student who had access to a scanner made a beautiful fractal map of Africa, using an IFS with eight transformations. I showed this to the class the second time I taught the seminar, and I think it inspired many in the class to become really involved with this project. Two of the best were a fractal Donald Duck and a cow's head. Several students, including the maker of the cow's head, worked extremely hard and had all the concepts right, but just couldn't get their picture to come out the way they wanted. They came to me for help, and after much scrutiny, I was able to find a trivial mistake that had messed up their picture. When their picture came out the way they had envisioned it, the gratitude that they showed convinced me that this was more than just a homework problem to them; it was something they really wanted to accomplish.

In order to create her fractal cow's head, my student Sarah Goode, who has cows on her family farm in Ohio, began by sketching her desired cow's head on graph paper. Then, using

4. Conclusion

freehand drawing, she tried to fill it in with shrunken versions of the same thing. Sarah used 19 mini-cow's heads. For each of these, she had to find the coordinates of the tips of the ears and bottom of the mouth, and use a computer program to find the coefficients of the affine transformation that would send the points of the big head to the corresponding points on each little head. Finally, she put all of these coefficients into a Fractint file. Fractint plots a sequence of thousands of points, each obtained by applying one of the 19 affine transformations to the previous point. The sequence of points fills up the desired cow's head in a fractal fashion. The top of Figure 1 shows Sarah's original cow drawing; the bottom its fractal version. The affine transformations to generate the cow are given in the Appendix.

4 Conclusion

For me, one of the best things in teaching fractals is that, more than with other courses, I speculate and learn while I am teaching. A lot of this has to do with fractal geometry's being an emerging subject, without a complete text. One of these topics about which I have shared my wonder with my classes is the relationship of the Feigenbaum bifurcation diagram and the Mandelbrot set. Wherever there is a baby bug in the proboscis of the M set, there corresponds a region of periodic behavior in the bifurcation diagram, usually of small period for the easily visible bugs. See Figure 2. But these baby bugs are dense along the proboscis of the M set, and that means periodic behavior must occur at a dense open set of values along the bifurcation diagram. All the dark area of the bifurcation diagram usually is described as chaotic, but it seems from this that much of it is periodic of high period. I don't know that my students appreciate this difference or these speculations, but I find it interesting to try to sort out such things.

In conclusion, teaching about fractals has been very important to me. I feel that I reach a broader spectrum of people through this than through other mathematical topics. There are only a few people about whom I can definitely say, "My teaching about fractals influenced this person's opinion about mathematics," but there are a lot of people for whom I think this probably is true.

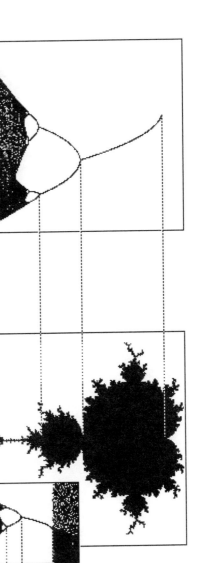

Figure 2: Top: the bifurcation diagram of $x^2 + c$, x and c real. Bottom: the Mandelbrot set. Note the relation between periodic windows in the bifurcation diagram and the baby bugs in the proboscis of the Mandelbrot set. The insert shows magnifications of corresponding parts of the bifurcation diagram and the Mandelbrot set.

Appendix

Representing an affine transformation by

$$T_i \begin{bmatrix} x \\ y \end{bmatrix} = s \begin{bmatrix} a_i & b_i \\ c_i & d_i \end{bmatrix} \begin{bmatrix} x \\ y \end{bmatrix} + \begin{bmatrix} e_i \\ f_i \end{bmatrix}.$$

here are the parameters, together with their probabilities, for the 19 transformations used to generate the cow.

a_i	b_i	c_i	d_i	e_i	f_i	prob
.1625	−.06839	.05	.16717	−3.5976	7.9331	.01
.3	0	0	.2584	−2.625	6.6467	.02
0	.3343	−.3	0	−3.7587	7.825	.02
.05	−.0608	.1	.0608	−.4757	8.9757	.01
.05	.0608	−.1	.0608	.4757	8.9757	.01
−.375	0	0	−.4407	0	9.4263	.04
0	−.3343	.3	0	3.7587	7.825	.02
−.3	0	0	.2584	3.7587	6.6467	.02
.1625	.06839	−.05	.1672	3.5976	7.4331	.01
.1	−.076	−.025	.1064	−.9696	5.9574	.17
−.1	.076	−.025	.1064	.9696	5.9574	.17
.45	0	0	.6505	0	.1398	.04
.0375	−.0532	.025	.0912	−.8537	.9635	.15
−.0375	.0532	.025	.0912	.8537	.9635	.15
.1625	.3419	−.28	.1945	−1.8868	1.0472	.03
.1625	−.3419	.28	.1945	1.8868	1.0472	.03
.4	0	0	.7903	−1.25	1.1839	.04
.4	0	0	.7903	1.25	1.1839	.04
−.25	0	0	−.304	0	3.2466	.02

Chapter 8

The Fractal Geometry of the Mandelbrot Set: Periods of the Bulbs

Robert L. Devaney

One of the most intricate and beautiful images in all of mathematics is the Mandelbrot set, discovered by Benoit Mandelbrot in 1980. Most people within the mathematics community, and many people outside of the discipline, have seen this image and have marveled at its geometric intricacy. Unfortunately, only a few of these people are acquainted with the equally beautiful mathematics that lurks behind this image.

In this chapter we present a few of these mathematical ideas in an elementary setting. All of these ideas were presented to high school students who participated in a Chaos Club organized by Jonathan Choate, Mary Corkery, Beverly Mawn, and the author at Boston Technical High School for several academic years. The goal of the club was to introduce inner city high school students to some of the beauty and excitement of contemporary mathematics. In this chapter I describe several of the computer experiments that were performed by the participating students.

1 The Chaos Club

The Chaos Club featured a number of activities that centered around chaotic dynamics and fractal geometry. Much of the time was spent understanding the *chaos game* and other iterated function systems. There are many worthwhile activities involving these notions that give students a taste of modern mathematics while linking nicely with what students are currently learning in their mathematics classes. For example, the chaos game involves the geometry of linear transformations in an essential way. This is one aspect of mathematics that is sorely lacking from the curriculum, but which is easy to teach using technology. Asking students to identify the attractors generated by various iterated function systems is an excellent means to introduce this subject. See Devaney (1995) for more details on these activities.

Regarding iteration, the students with whom we worked had one distinct advantage—they all knew how to use a spreadsheet before joining the club. A part of their computer literacy course is a requirement that before graduation all Boston Public High School students know how to use a spreadsheet. It is unfortunate that the mathematics department does not take advantage of this unusual opportunity, since the spreadsheet is an ideal tool with which to teach and to explore certain mathematical concepts. For example, the basic iterative processes described in the next section can easily be carried out on a spreadsheet. Of course, graphing calculators such as the *TI-83* also offer this benefit, but one can display just about all the geometric data such as time series and histograms described below easily and effectively with a spreadsheet. In the Chaos Club we used spreadsheets to introduce simple quadratic iterations of both a real and a complex variable.

In this chapter we describe the activities and mathematics related to the Mandelbrot set. In the Chaos Club, students used the computer to draw the world's largest copy of the Mandelbrot set (well, maybe not quite). Students were assigned various bulbs or decorations to compute, and the entire image was assembled on a wall of a classroom. Then the students were asked to understand the mathematics behind this image—the periods of the bulbs. In this chapter, we describe both the mathematical skills that the students learned as well as the computer experiments that they performed.

2 Iteration

The Mandelbrot set is generated by iteration. Iteration means to repeat a process over and over again. In mathematics this process is most often the application of a mathematical function. For the Mandelbrot set, the function involved is the simplest nonlinear function imaginable, namely $x^2 + c$, where c is a constant. As we go along, we will specify exactly what value c takes.

To iterate $x^2 + c$, we begin with a *seed* for the iteration. This is a (real or complex) number denoted by x_0. Applying the function $x^2 + c$ to x_0 yields the new number

$$x_1 = x_0^2 + c$$

Now, we iterate, using the result of the previous computation as the input for the next. That is

$$x_2 = x_1^2 + c$$
$$x_3 = x_2^2 + c$$
$$x_4 = x_3^2 + c$$

and so forth. The list of numbers $x_0, x_1, x_2, x_3, \ldots$ generated by this iteration has a name, the *orbit* of x_0 under iteration of $x^2 + c$. One of the principal questions in this area of mathematics is: What is the fate of typical orbits? Do they converge or diverge? Do they cycle or behave erratically? In a real sense, the Mandelbrot set is a geometric version of the answer to this question.

Let's begin with a few examples. Suppose we start with the constant $c = 1$. Then, if we choose the seed 0, the orbit is

$$x_0 = 0$$
$$x_1 = 0^2 + 1 = 1$$
$$x_2 = 1^2 + 1 = 2$$
$$x_3 = 5$$
$$x_4 = 26$$
$$x_5 = \text{big}$$
$$x_6 = \text{bigger}$$
$$\vdots$$

and we see that this orbit tends to infinity.

As another example, for $c = 0$, the orbit of the seed 0 is quite different:

$$x_0 = 0$$
$$x_1 = 0$$
$$x_2 = 0$$
$$\vdots$$

This orbit remains *fixed* for all iterations.

If we now choose $c = -1$, something else happens. For the seed 0, the orbit is

$$x_0 = 0$$
$$x_1 = -1$$
$$x_2 = 0$$
$$x_3 = -1$$
$$\vdots$$

Here we see that the orbit bounces back and forth between 0 and -1, a *cycle of period 2*.

To understand the fate of orbits, it is often easiest to proceed geometrically. Accordingly, a *time series* plot of an orbit may give more information about the fate of that orbit. In Figures 1a–1d, we have displayed the time series for $x^2 + c$ where $c = -1.1, -1.3, -1.38$, and -1.9. In each case we have computed the orbit of 0. Note that the fate of the orbit changes with c. For $c = -1.1$, we see that the orbit approaches a 2-cycle. For $c = -1.3$, the orbit tends to a 4-cycle. For $c = -1.38$, we see an 8-cycle. And when $c = -1.9$, there is no apparent pattern for the orbit; mathematicians use the word *chaos* for this phenomenon. To see this in another light, in Figure 2 we have plotted a histogram of the first 20,000 points on the orbit of 0 under $x^2 - 1.9$. In this picture the interval $-2 \leq x \leq 2$ is subdivided into 400 subintervals. The histogram was incremented by one unit each time the orbit entered one of these subintervals.

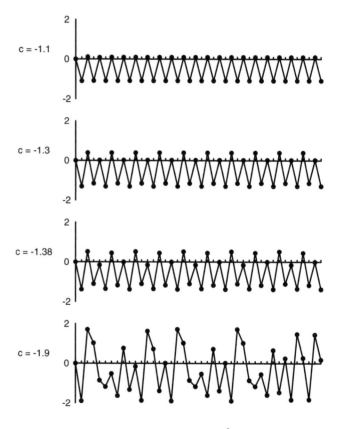

Figure 1: Time series for $x^2 + c$.

Before proceeding, let us make a seemingly obvious and uninspiring observation. Under iteration of $x^2 + c$, either the orbit of 0 goes to infinity, or it does not. When the orbit does not go to infinity, it may behave in a variety of ways. It may be fixed or cyclic or behave chaotically, but the fundamental observation is that there is a dichotomy: Sometimes the orbit goes to infinity, other times, it does not. The Mandelbrot set is

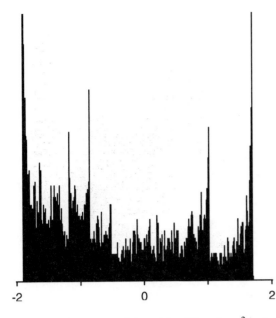

Figure 2: Histogram of the orbit of 0 under $x^2 - 1.9$.

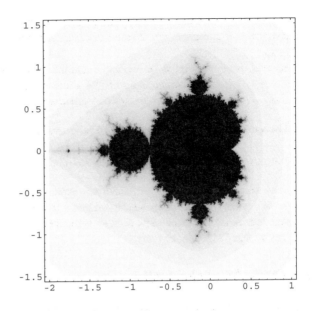

Figure 3: The Mandelbrot set.

a picture of precisely this dichotomy in the special case where 0 is used as the seed. Thus the Mandelbrot set is a record of the fate of the orbit of 0 under iteration of $x^2 + c$.

How then is the Mandelbrot set a planar picture? The answer is, instead of considering real values of c, we allow c to be a complex number. For example, the orbit of 0 under $x^2 + i$ is given by

$$x_0 = 0$$
$$x_1 = i$$
$$x_2 = -1 + i$$
$$x_3 = -i$$
$$x_4 = -1 + i$$
$$x_5 = -i$$
$$\vdots$$

and we see that this orbit eventually cycles with period 2. If we change c to $2i$, then the orbit behaves very differently

$$x_0 = 0$$
$$x_1 = 2i$$
$$x_2 = -4 + 2i$$
$$x_3 = 12 - 14i$$
$$x_4 = 52 - 334i$$
$$\vdots$$

and we see that this orbit tends to infinity in the complex plane (the numbers comprising the orbit recede farther and farther from the origin). Again we make the fundamental observation either the orbit of 0 under $x^2 + c$ tends to infinity, or it does not.

3   The Mandelbrot set

The Mandelbrot set puts some geometry into the fundamental observation above. The precise definition is: The Mandelbrot set $\mathcal{M}$ consists of all of those (complex) c-values for which the corresponding orbit of 0 under $x^2 + c$ does not escape to infinity. From our previous calculations, it follows that $c = 0, -1, -1.1, -1.3, -1.38$, and i all lie in the Mandelbrot set, whereas $c = 1$ and $c = 2i$ do not.

At this point, a natural question is: Why would anyone care about the fate of the orbit of 0 under $x^2 + c$? Why not the orbit of i? Or $\pi + 3i$? Or any other complex seeds, for that matter? As we shall see below, there is a very good reason for inquiring about the fate of the orbit of 0; somehow the orbit of 0 tells us a tremendous amount about the fate of other orbits under $x^2 + c$.

Before focusing on this idea, note that the very definition of the Mandelbrot set gives us an algorithm for computing it. We simply consider a square in the complex plane (usually centered at the origin with sides of length 4). We overlay a grid of equally spaced points in this square. Each of these points is to be considered a complex c-value. Then, for each such c, we ask the computer to check whether the corresponding orbit of 0 goes to infinity (escapes) or does not go to infinity (remains bounded). In the former case, we leave the corresponding c-value (pixel) white. In the latter case, we paint the c-value black. Thus the black points in Figure 3 represent the Mandelbrot set.

Two points need to be made. Figure 3 is only an approximation of the Mandelbrot set. Indeed, it is not possible to determine whether certain c-values lie in the Mandelbrot set. We can only iterate a finite number of times to determine if a point lies in $\mathcal{M}$. Certain c-values close to the boundary of $\mathcal{M}$ have orbits that escape only after a very large number of iterations.

A second question is: How do we know that the orbit of 0 under $x^2 + c$ really does escape to infinity? Fortunately, there is an easy criterion which helps answer this question:

The Escape Criterion: Suppose $|c| \leq 2$. If the orbit of 0 under $x^2 + c$ ever lands outside of the circle of radius 2 centered at the origin, then this orbit definitely tends to infinity.

It may seem that this criterion is not too valuable, as it only works when $|c| \leq 2$. However, it is known that the entire Mandelbrot set lies inside this disk, so these are the only c-values we need consider anyway.

4 Periods of the bulbs

Note that the Mandelbrot set consists of many small decorations. Closer inspection of these decorations shows that all of them are different in shape. See Figure 4.

For example, consider any decoration directly attached to the main cardioid in $\mathcal{M}$. We call this bulb a *primary* bulb or decoration. Attached to this decoration in turn are infinitely many smaller decorations, as well as what appear to be antennas. In particular, as is clearly visible in Figure 4, the *main antenna* attached to each decoration seems to consist of a number of spokes that varies from decoration to decoration.

There is a beautiful relationship between the number of spokes on these antennas and the dynamics of $x^2 + c$ for c inside the primary bulb. It is known that if c lies in the interior of such a decoration, then the orbit of 0 is attracted to a cycle of a given period n. The number n is the same for any c inside this main decoration. (It is a multiple of n for c inside the other smaller decorations attached to the primary decoration.) For example, $c = -1$ and $c = -1.1$ both lie inside the largest primary bulb just to the left of the main cardioid. For these c-values, the orbit of 0 is attracted to a cycle of period 2.

Before discussing the number of spokes, it is useful to perform a series of computer experiments that yield the periods of some of the other primary decorations. You can easily check the following facts using a computer and any of a number of available software packages. (See, for example Georges, Johnson & Devaney (s1992).)

point c in bulb	period of bulb
$-0.12 + 0.75i$	3
$-0.5\ \ + 0.56i$	5
$0.28 + 0.54i$	4
$0.38 + 0.333i$	5
$-0.62 + 0.43i$	7
$-0.36 + 0.62i$	8
$-0.67 + 0.34i$	9
$0.39 + 0.22i$	6

Figure 4: Several decorations on the Mandelbrot set.

It is easy to check that $\mathcal{M}$ is symmetric about the real axis and that these periods hold for the complex conjugate c-values. This gives us a larger table of results. In Figure 5 we have summarized some of these results graphically.

Further experimentation reveals the remarkable relationship we mentioned between the number of spokes in the largest antenna attached to a primary decoration and the period of that decoration. These numbers are exactly the same! (Don't forget to count the spoke emanating from the primary decoration to the main junction point.) See Figure 6.

5 Julia sets

There is a second, more dynamic way to calculate the periods of these primary bulbs in $\mathcal{M}$. To explain this, we have to introduce the notion of a Julia set.

The Julia set for $x^2 + c$ is subtly different from the Mandelbrot set. For $\mathcal{M}$, we calculated only the orbit of 0 for each

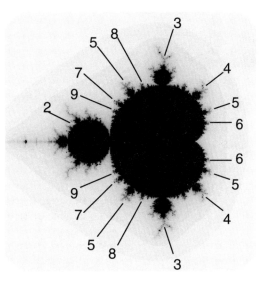

Figure 5: Periods of the primary bulbs in $\mathcal{M}$.

Period 3 bulb

Period 4 bulb

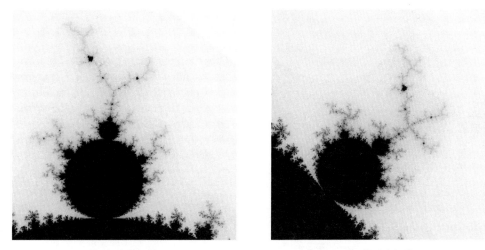

Period 5 bulb

Period 7 bulb

Figure 6: Note that the period of the bulb is the same as the number of spokes in the antenna.

$$x_3 = r_0^8 e^{i(8\theta)}$$

$$\vdots$$

$$x_n = r_0^{2^n} e^{i(2^n \theta)}$$

Thus, the orbit of x_0 tends to infinity if $r_0 > 1$ because

$$\lim_{n \to \infty} r_0^{2^n} = \infty.$$

Because

$$\lim_{n \to \infty} r_0^{2^n} = 0,$$

when $r_0 < 1$, it follows that any seed inside the circle of radius 1 centered at the origin tends to the origin (and so does not escape).

Finally, if $r_0 = 1$, the orbit of x_0 remains on the unit circle for all iterations (and therefore does not escape). It follows that J_0 consists of all those seeds whose orbits lie on or inside the unit circle centered at the origin.

Incidentally, the fate of orbits of x^2 that lie on the unit circle is quite an interesting story. These are precisely the orbits that behave in a chaotic fashion. See Devaney (1989) for more details.

Other Julia sets are much more difficult to compute. To see them, we must use a computer. The algorithm is, of course, a direct consequence of the definition of J_c. We simply consider a grid of points centered at the origin and compute the orbit of each of these points under $x^2 + c$. Either this orbit tends to infinity (in which case the seed was not in J_c) or else the orbit does not (and so the seed lies in J_c).

Figure 7: Julia sets for several c-values.

Important Difference

The **Mandelbrot set:**

- is a picture in parameter space

- records the fate of the orbit of 0

The **Julia set:**

- is a picture in dynamical plane

- records the fate of all orbits

---

c-value and then displayed the result. A c-value lies in $\mathcal{M}$ if the corresponding orbit of 0 does not escape to infinity. Thus, $\mathcal{M}$ is a picture in the c-plane, the parameter plane.

For the Julia sets, we fix a c-value and then consider the fate of all possible seeds for that fixed value of c. Those seeds whose orbits do not escape form the *Julia set* of $x^2 + c$. Orbits that do escape do not lie in the Julia set. Thus we get a different Julia set of each different choice of c. That is, the Julia set is a picture in the dynamical plane, not the parameter plane. We denote the Julia set for $x^2 + c$ by J_c. In Figure 7 we have displayed the Julia sets for a variety of c-values.

There is one c-value for which the Julia set is easy to compute by hand, namely $c = 0$. The Julia set J_0 for $x^2 + 0$, is easily seen to be the closed unit disk centered at the origin of the plane. The reason is simple. Let x_0 be a seed. In polar coordinates we may write $x_0 = r_0 e^{i\theta}$. But then the orbit of x_0 under x^2 is

$$x_1 = r_0^2 e^{i(2\theta)}$$
$$x_2 = r_0^4 e^{i(4\theta)}$$

6   The fundamental dichotomy

One of the most beautiful results in all of complex dynamics dates back to 1919 and was proved by both G. Julia and P. Fatou. The *fundamental dichotomy* for $x^2 + c$ is this. For each c-value, the Julia set is either a connected set or a Cantor set. A connected set is a set that consists of just one piece. It may be a simple object, like the unit disk—the Julia set for x^2—or it may be any of the more complicated figures depicted in Figure 7. All of these Julia sets are connected.

On the other hand, a Cantor set consists of infinitely many pieces—in fact, uncountably many distinct pieces. Moreover, each piece is a point and every point in the Cantor set is a limit point of other points in this set. So a Cantor set should be visualized as a cloud of points—no two points touching, but infinitely many points scattered in any region around a given point.

So the fundamental dichotomy says that Julia sets for x^2+c come in one of two varieties: connected sets (one piece) or Cantor sets (infinitely many pieces). There is no in-between; there are no c-values for which J_c consists of 10 or 20 or 756 pieces.

How do we decide what shape a given J_c assumes? Amazingly, it is the orbit of 0 that determines this. If the orbit of 0 tends to infinity under iteration of $x^2 + c$, then that J_c is a Cantor set. On the other hand, if the orbit of 0 does not tend to infinity, that J_c is a connected set.

A visual way to view this dichotomy is given by the Mandelbrot set. If c lies in $\mathcal{M}$, then we know that the orbit of 0 does not escape to infinity under iteration of $x^2 + c$, so J_c must be connected. If c does not lie in $\mathcal{M}$, then J_c is a Cantor set. This dichotomy thus gives us a second interpretation of the Mandelbrot set.

The Mandelbrot set consists of all c-values for which

- J_c is connected, or, equivalently

- the orbit of 0 under $x^2 + c$ does not tend to infinity.

It is amazing that the orbit of 0 knows the shape of the Julia set for x^2+c. By a theorem of Fatou and Julia, the reason that 0 is so special stems from the fact that 0 is the critical point of $x^2 + c$. (The derivative of $x^2 + c$ is $2x$, and this derivative only vanishes at $x = 0$.)

7   Back to $\mathcal{M}$

Besides the period, there is another way to attach an integer to each primary decoration in $\mathcal{M}$. In Figure 8 we have displayed the Julia set for $c = -0.12 + 0.7i$. This Julia set is often called Douady's rabbit. Note that the image looks like a fractal rabbit. The rabbit has a main body with two ears attached. But everywhere you look you see other pairs of ears.

Another way to say this is that the Julia set contains infinitely many junction points at which three distinct black regions in J_c are attached. In Figure 9 we have magnified a portion of the fractal rabbit to illustrate this.

The fact that each junction point in this Julia set has three pieces attached to it is no surprise, since this c-value lies in a primary period 3 bulb in the Mandelbrot set. This is another fascinating fact about $\mathcal{M}$. If you choose a c-value from one of the primary decorations in $\mathcal{M}$, then, first of all, J_c must

Figure 8: The fractal rabbit.

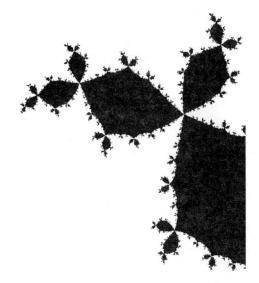

Figure 9: A magnification of the fractal rabbit.

be a connected set, and second, J_c contains infinitely many special junction points. Each of these points has exactly n regions attached to it, where n is exactly the period of the bulb. Figure 10 illustrates this for periods 4 and 5.

8   Classroom activities

One of the principal goals of the Chaos Club was to acquaint students with different representations of the same idea. Iteration provides a wonderful opportunity to do this. During the first weeks of the club, students became familiar with the notion of iteration by listing orbits, plotting histograms, and drawing time series. Working only with $x^2 + c$ for real values of x and c, students were exposed to such contemporary mathematical phenomena as chaos and bifurcations.

One image that unifies all of these ideas is the orbit or bifurcation diagram. While we did not have time in the Chaos Club to investigate this image fully, the relationship between

Figure 10: Period 4 and 5 Julia sets.

this picture and the Mandelbrot set is described in Devaney (1991). It is worthwhile trying to get students to appreciate the relationship between these two dissimilar images.

A large portion of time in the club was devoted to explaining the Mandelbrot set. Students used the software package Georges, Johnson & Devaney (s1992) to understand the meaning of the Mandelbrot set as the set of parameter values for which the orbit of 0 does not escape. Of course, the idea of a parameter space is one that is quite unfamiliar to students. Nonetheless, with software that displays the fate of the orbit of 0 as a mouse is dragged over the Mandelbrot set, students quickly catch on.

One of the main experiments that the students performed was aimed at discovering the relationship between the geometry and the periods of the bulbs. Working in groups, students first magnified, then printed, various primary bulbs. The printouts were assembled in a large mural depicting the Mandelbrot set together with the associated periods. This large view enabled students to see without difficulty the fact that the period was equal to the number of spokes on the principal antenna. This large image still hangs in the school's computer lab and is a constant source of many questions from current students.

Once the students understood the mathematics behind the periods of the bulbs, they naturally began looking for the pattern that governed the arrangement of the bulbs. This re-

quires further work with the quadratic Julia sets and is described in Devaney (1999).

9 Summary

The Mandelbrot set possesses an extraordinary amount of structure. We can use the geometry of $\mathcal{M}$ to understand the dynamics of $x^2 + c$. Or we can take dynamical information and use it to understand the shape of $\mathcal{M}$. This interplay between dynamics and geometry is, on the one hand, fascinating and, on the other, still not completely understood. Much of this interplay has been catalogued in recent years by mathematicians such as Douady, Hubbard, Yoccoz, McMullen, and others, but much more remains to be discovered.

For more details on the Mandelbrot and Julia sets, we refer to the American Mathematical Society volumes Devaney & Keen (1989) and Devaney (1994). References Devaney (1989) and Devaney (1994) contain more general referencs to the theory of dynamical systems.

Note: An interactive version of this paper is available on the worldwide web at

```
http://math.bu.edu/DYSYS/dysys.html
```

Comments are most welcome—send to

```
bob@bu.edu
```

Chapter 9

Fractals—Energizing the Mathematics Classroom

Vicki G. Fegers and Mary Beth Johnson

1 Introduction

Chaos and fractals—they are a part of everyday conversation in the mathematics world today, but it wasn't so long ago that fractals were unknown to most mathematics teachers. The introduction to fractals has sometimes been a slow process for the mathematics teacher, but once the introduction is made, the beauty of fractals inspires educators to learn more. The journey that led us to fractals is different for each of us and can best be described individually.

1.1 Mary Beth's narrative

The first time I remember seeing the word "fractal" was when I read a magazine article in the late 1980s. I cut out the article, posted it on my classroom bulletin board, and answered any questions my students had about Fractal Geometry to the best of my ability (which at that time was limited to the information in the article!). Flash forward to the National Council of Teachers of Mathematics (NCTM), a national conference in New Orleans in 1991. There I sat, mesmerized by a speaker who had a ballroom full of mathematics teachers sitting on the edge of their seats, hanging on his every word, as he shared his knowledge and love of chaos and fractals with a room full of strangers. Dr. Heinz-Otto Peitgen's enthusiasm for this new world of mathematics left thousands of teachers wanting more.

I was enthralled, but upon my return to the classroom did not know where to go with this newfound knowledge. Thinking that a little bit of knowledge can be a dangerous thing, I did not venture far beyond the walls of my classroom with what I had learned. However my friend Vicki had other ideas. She had decided to pursue her masters degree and enrolled in the Master of Science in Teaching (MST) program at Florida Atlantic University. The next thing I knew, she was in her second semester and eagerly attending a fractal class taught by none other than H.-O. Peitgen. The following year she recruited a mutual friend to enroll in Peitgen's spring semester class. During the 1993 NCTM convention in Seattle, the two of them reserved front row seats for Peitgen's featured lecture and introduced me to him before he spoke, telling him that I would be in his class next year before I had even considered the possibility. It didn't take much to convince me. And what an amazing experience! I was hooked from the very first day when he spoke of the beauty of mathematics. Mathematics beautiful? I loved mathematics for the challenge it had always offered me, but I never viewed it as being beautiful, nor had I ever had a teacher show such interest in the learning process at the university level. H.-O. Peitgen had a genuine passion for his subject and would often stop to look at us to see if we were comprehending the pearls of wisdom he was offering us. If he saw just one puzzled look, he would patiently recount his steps until the light bulb of insight shone in all faces. He taught me more than fractals that semester—he taught me how to be a better teacher.

1.2 Vicki's narrative

My adventures with fractals began approximately a decade ago. Mathematics teachers at the school where I was teaching started talking about fractals. "Say, Vicki, do you know anything about these things called fractals?" Of course, I had to say no, but the seed was planted. A couple of months later, another teacher showed me the Sierpinski Triangle and asked if I knew anything about it. I had to say no again, but I was determined to learn more about it. As luck would have it, soon I was able to attend the NCTM conference in New Orleans where Heinz-Otto Peitgen was the speaker at a keynote session. POW!!—I was hooked!! I had to learn more.

As it turns out, later that summer Peitgen was hired by Florida Atlantic University where I was just beginning a masters program. Peitgen was to teach a class for the program each spring semester. Was I thrilled!! It was so incredible to me that I would receive instruction in chaos and fractals from one of the world's masters. From that first class, I knew that I would never learn enough. Taking a class taught by H.-O. Peitgen is more a journey than an event, a journey through the fascinating world of fractals. The destination was never evident, but, by the end of the course a map of the journey had been drawn. All the connections had been made—I knew exactly why we had taken that path.

For our evaluation, we were asked to keep a journal of our thoughts and reactions to what we were learning. Some classmates and I approached Dr. Peitgen about doing something a little different. Rather than keeping a journal, he approved our proposal to develop lessons, to be used by students in the classroom. The highly theoretical material he presented in class was written at a very basic level that maintained the essence of his course work but could be understood by elementary to high school students. The lessons we created sparked Peitgen's interest, beginning a closer association with him as he involved us in the process of writing a grant for the National Science Foundation.

1.3   The grant

In 1995, H.-O. Peitgen was awarded a three year National Science Foundation grant, *Mathematics and Science Teachers Enhancement through Chaos and Fractals*. The grant was a collaborative project between FAU and the School Board of Broward County, Florida.

The grant philosophy was inspired by the principle of learning by exploring, and by the exciting areas of chaos theory and fractal geometry. Fractal geometry is a new language used to describe and analyze the complex forms found in nature. Chaos theory gives a way of seeing order underlying essentially unpredictable phenomena. Both chaos theory and fractal geometry are new, visual, relevant to many disciplines, very naturally lend themselves to computer supported activities, and can be understood (at some level) by students with relatively little mathematical background.

The principal goal of the project was to shape teachers' mental images of mathematics and science by having them experience creative mathematical and scientific activities centered around recognizing and solving problems. These activities served to increase teachers' and students' capacities to confront novel problems and to increase their initiative for self-directed learning and collaborative group work. The project also supported instruction in middle and high school mathematics of centrally important topics, including patterns, scaling, transformations, symmetry, self-similarity, and geometry.

In the spring of 1995, an invitation was extended to science and mathematics teachers of Broward County to participate in the three year project as Lead Teachers. Twenty teachers were selected to spend three weeks in the summer of 1995 with some of fractal geometry's and chaos theory's great mathematicians and scientists—Dr. Heinz-Otto Peitgen, Dr. Richard Voss, Dr. Evan Maletsky, Dr. Terry Perciante, and Dr. Robert L. Devaney. After two weeks of intense training, the twenty Lead Teachers separated into five groups to begin writing the first iteration of the CATEs—Curriculum and Textbook Enhancements—*Fractal Dimension, Diffusion-Limited Aggregation: Fractal Growth, A Connection between Sierpinski's Triangle and Pascal's Triangle, Lessons in Paper Folding*, and *Fractal Wallhanging*. A CATE is a small cluster of activities designed to explore a mathematical or scientific idea drawn from chaos and fractals. It constitutes a short phase of explorative learning that can be used to enhance a topic in the curriculum or to connect a topic with other parts of the curriculum. CATEs involve looking at open-ended problems, looking at a problem from different points of view, and building models. CATEs consistently employ hands-on activities and manipulatives, and are designed to have students work together in cooperative groups.

During the 1995–96 school year, the Lead Teachers met monthly to discuss the first draft of the CATEs and share classroom experiences in preparation for the upcoming summer institute. In the spring of 1996, each group met separately with Peitgen to work on improving the content of each particular CATE and the second iteration was ready for the first *Mathematics and Science Teacher Enhancement through Chaos and Fractals* institute.

1.4   The summer institutes

During the summer of 1996, eighty middle and high school mathematics and science teachers attended the three week summer institute. Teachers attended the institute in teams in order to support one another during the school year as they implemented what they learned during the institute. Participants spent the first four days of the week attending two lectures in the morning presented by the Institute Instructors: H.-O. Peitgen, R. F. Voss, E. Maletsky, and T. Perciante. In the afternoon, participants divided into groups and rotated through two seventy-five minute workshops on the CATE activities directed by the Lead Teachers. Recognizing that computer and graphing calculator technology plays a vital part of the study of chaos and fractals, workshops in these two areas were also provided for the participants. The fifth day was reserved for lectures only. Part of the requirements for the participants, in lieu of a post test, was to keep a daily journal of their experiences in the lectures and workshops. Journals were collected and read by the Lead Teachers on a weekly basis. It was a learning experience for all, Lead Teachers and participants alike.

During the follow-up monthly weekend meetings throughout the 1996–97 school year, the participants shared the activities they had tried in their classrooms and brought in samples of student work. They participated in chaos writing activities, random walks, computer labs, and received additional TI-82 calculator instruction. Recognizing the need for another science instructor, Dr. Larry Liebovitch was invited to join the Institute Instructors as a weekend lecturer. His contributions in the areas of fractal biology and brain wave patterns proved to be overwhelmingly popular.

During the spring of 1997, preparations were made for the second summer institute. From the experiences gained during the first summer institute, several changes were made to the structure of the institute. Participants were grouped according to the level of students they taught, i.e., all middle school teachers were kept together and all high school teachers were kept together. Teachers from the same school were assigned to the same group, to allow ongoing conversations in planning the implementation of the CATEs at their schools. Not only were the contents of the courses tailored for the appropriate level group, in addition the CATEs were presented to the groups at a level appropriate for the learners they would be teaching.

Journal guidelines were developed and distributed the first day. Journals were collected once a week on Thursday. Each Lead Teacher was assigned to read 4-5 journals. On Friday, the Lead Teachers met with their Journal Mentoring groups to facilitate discussions about questions or topics raised in the journals, provide verbal feedback to the participants, and serve as a clearinghouse for further questions or concerns. This made the institute a much more personal experience.

Due to the important role of the calculator, graphing calculators were updated from TI-82 to TI-83 and a separate calculator CATE was added to the original five CATEs. Reimund Albers, a mathematics teacher from Bremen, Germany, was instrumental in the development of the calculator CATE. The many calculator programs he created allowed participants to explore the ideas presented in the other CATEs. The Chaos Game, introduced in the lectures and in the calculator CATE, was explored in more detail in a separate mini-session. Two new Lead Teachers were selected from the first summer participants to replace Lead Teachers who had moved out of the district. L. Liebovitch and Dr. Jim Brewer joined the Institute Instructors on a full time basis, adding enthusiastically received science lectures and a series of lectures connecting chaos and probability.

The biggest change was in the CATEs themselves. During the spring of 1997, Lead Teachers did a major rewrite of the CATEs. Mary Beth and Vicki were assigned the task of overseeing the rewriting of the CATEs in general, and specifically, Mary Beth rewrote *Lessons in Paper Folding* and Vicki rewrote *Fractal Wallhanging*. The Teacher Guidelines were improved with the addition of concise and detailed mathematics and science explanations. Annotated illustrations and graphics provided clear directions for classroom implementation. A major focus was student activities supported by separate student sheets, along with solutions and possible extensions. The intent of the third iteration of the CATEs was for teachers to be able to use the document without attending the institute. The daily and weekly schedules were changed from the previous year. Each day, the participants attended three lectures and two CATE sessions. This schedule was followed four days a week. On Friday only two lectures were given, allowing the Lead Teachers time to meet their Journal Mentoring groups in the afternoon.

The success of the summer institutes verified that fractals and chaos have captured the attention, enthusiasm, and interest of many people. To the casual observer, their color, beauty, and geometric structure captivates the visual senses like few other things ever experienced in mathematics. To the student, fractals and chaos bring mathematics out of the past and into the twenty-first century. To the teacher, fractals and chaos offer a unique opportunity to illustrate both dynamics of mathematics and its many connecting links to science and technology.

The following sections present a sample of the CATEs *Lessons in Paper Folding* and *Fractal Wallhanging*.

2 Lessons in paper folding

Origami is the ancient art of paper folding. The modern version of paper folding addressed in the CATE sampled here integrates the beauty of mathematics with the physical act of folding paper. Mathematical patterns of paper folding can be used to make predictions about the process far beyond the physical limitations of physical paper folding. Using *Scaled*, a TI-83 program written for this CATE and presented in Appendix 8, students can create scaled paper folding curves through Step 10.

2.1 Unit 1. Symmetric paper folding

The study of patterns is a mathematical concept that invites students to utilize higher order thinking skills. Paper folding introduces patterns in a way that leads to the discovery of formulas and sequences. The activities in this unit are designed to be done in cooperative groups of two or four students.

Grade level: 5–12
Materials: 5–6 strips of paper ($1'' \times 11''$) per student
Transparency #1 (Appendix 1)
Overhead projector
Activity Sheet 1 (Appendix 3, answers in Appendix 5)

Notes to the teacher

Beginning with a one inch wide strip of paper, mark a starting point on the right edge of the strip as illustrated. Fold the left

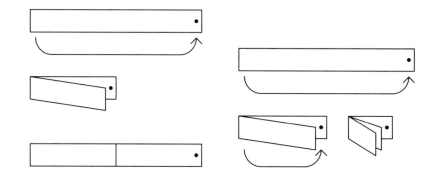

Figure 1: Left: Making a + fold. Right: Making a ++ fold.

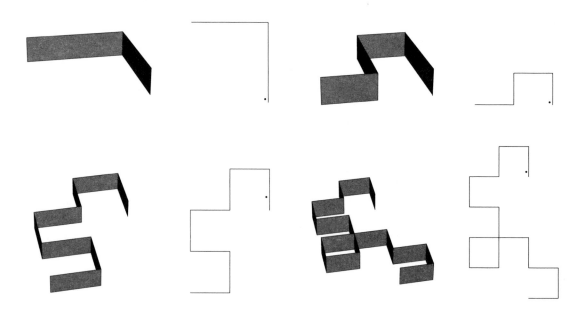

Figure 2: Top: Steps 1 through 4 of the systematic paperfolding process, perspective and overhead views.

edge of the paper strip over the right. Match the edges and crease the fold. **This method of folding will be referred to as making a + fold.** See the left side of Figure 1. Unfold the paper strip and observe it has one crease that divides the strip into two segments.

Place the unfolded paper strip on the overhead projector so that it forms a right angle at the crease; when following the direction of the strip from the mark, the strip begins by pointing north and ends by pointing west. The resulting figure is referred to as **Step 1** of the systematic paper folding process. See Figure 2.

Starting at the marked edge, travel along the paper and record the direction, *L* or *R*, of the turn at the crease. In all steps of this exercise, have students compare answers. Resolve any disagreements before proceeding. Illustrate the correct placement by putting the folded strip on the overhead projector.

Using two new strips of paper, mark a starting point on the right edge, make a **+ + fold** with one strip and a **+ + + fold** with the other. These produce **Step 2** and **Step 3** of the systematic paper folding process.

Starting at the marked edge, travel along the paper and record the direction of the turn at each crease. Note **Step 2** has turn directions *LLR*; these three creases divide the strip into four segments. **Step 3** has turn directions *LLRLLRR*; these seven creases divide the strip into eight segments.

With a fourth strip of paper, make a **+ + + + fold**. The resulting figure is **Step 4** of the systematic paper folding process. The turn directions are *LLRLLRRLLLRRLRR*.

It should be obvious at this point that the figure is becoming more complex. What is different about the result of this step? *The figure appears to curve around to touch itself for the first time, creating an enclosed area.* In the remainder of this unit, the resulting figures will be referred to as **curves**.

An effective way to introduce mathematical patterns is the relationship between the number of folds, the resulting number of creases in the paper strip, and the number of segments formed. It is important for students to use words to describe the patterns they observe. The students might say, "The number of segments doubles with each additional fold," and "The number of creases is always one less than the number of segments." Listen carefully to ensure that the students demonstrate an understanding of the patterns before leading them into the discovery of the formulas for the patterns. Even when students can correctly describe the process, writing formulas for the number of segments and creases is a difficult process for them. Transparency #1 (Appendix 1) is provided to assist the teacher and the students in developing the formulas for the number of segments and creases.

Special notes

The activities in Unit I combine two of the units of the Paper Folding CATE. It is also interesting to approach the units separately by first folding the strips of paper and examining the mathematics in the number of folds, segments, and creases before having the students discover the need for a common orientation in the placement of the strips of paper. Once the limit of physical folding paper is reached, the students are motivated to see the necessity of a mathematical coding system.

Unit 1 offers students the opportunity to investigate patterns through the process of folding paper. Depending on the level of the student, the teacher can use paper folding to introduce sequences (the number of segments formed—a geometric sequence), to discover the formulas for generating the number of segments and creases formed as they relate to the number of folds made, and to look for patterns in the left and right turns. An extension of this unit can be in its application to problems found in Activity Sheet 1 (Appendix 3).

2.2 Unit 2. The Inflation Law

Any step of the paper folding process can be represented as a sequence of L and R turns, the **code** of the step. Ask the students, "How many steps can be created by folding paper?" (physically paper can only be folded about seven times) and, "Can the code of any step be determined without folding paper?" Students usually equate the number of steps to the length of the original strip of paper—the longer the paper strip, the more steps that can be folded. These simple questions generate interesting discussion and discovery on the part of the students. Eventually the students will realize that it will be physically impossible to continue folding paper and determine the code for subsequent steps. This unit will explore a way to generate the paper folding coding sequence without actually folding paper.

Grade level: 5–12
Materials: 5–6 strips of paper ($1'' \times 11''$) per student
Transparency #2 (Appendix 2)
Overhead projector
Activity Sheet 2 (Appendix 4, answers in Appendix 6)

Notes to the teacher

Beginning with a strip of paper, mark a starting point on the right edge. Make a + fold. Unfold the paper. Place the paper on the desk with the mark on the left as illustrated in the top of Figure 3.

The mark has been used to represent the starting point of the strip of paper. In the English language, the orientation for reading and writing is to start on the left and proceed to the right. The code for each step will eventually be recorded on the paper strip; hence the reason for the placement of the mark on the left. Mark the crease with a number 1. Illustrate the process by revealing the first strip of transparency #2 (Appendix 2), illustrated in the bottom of Figure 3.

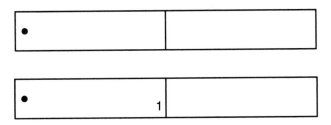

Figure 3: The beginning of the Inflation Law experiment.

With a second strip of paper, mark a starting point on the right edge and make a ++ fold. Unfold the paper after the first fold and with the mark on the left, mark the crease with a number 1. Refold the paper and make the second fold. Unfold the paper and mark the new creases with a number 2. Reveal the second strip of transparency #2, illustrated in the top of Figure 4.

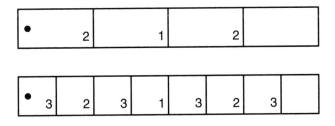

Figure 4: A ++ fold (top) and + + + fold (bottom), both with crease numbers marked.

With a third strip of paper, mark a starting point on the right edge and make a +++ fold and mark the creases accordingly. Reveal the third strip of transparency #2, illustrated in the bottom of Figure 4.

Instruct the students to write the code determined from the Unit 1 activites for each step on the strips of paper. Use the remainder of transparency #2 to illustrate to the students (Figure 5).

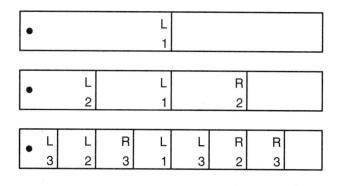

Figure 5: The +, ++, and + + + folds, with fold directions and crease numbers marked.

At the beginning of the activity, the original strip did not have a crease. In the + fold, one crease was added. In the ++ fold, two new creases were added. In the +++ fold, four new creases were added. Examining only the code of each new **additional crease** in each step, results in the following chart:

Step 1:   L
Step 2:   L   R
Step 3: L   R   L   R

Ask the students if they can discern a pattern. The new creases always appear on either side of a former crease and always code in an *LR* sequence. By taking any step in the paper folding sequence and inserting the *LR* sequence between each letter of the old code, the coding for the next step of the sequence can be found. Use the code for a Step 3 curve and insert the *LR* sequence.

Step 3 code: L   L   R   L   L   R   R
Insert LR sequence: L   R   L   R   L   R   L   R

The result is the code for a Step 4 curve.

LLRLLRRLLLRRLRR

This process of inserting an alternating *L* and *R* into the coding sequence of a step is called the *Inflation Law*. The Inflation Law inflates the coding of one step in the paper folding sequence to produce the coding of the next step without having to fold paper.

The Inflation Law can be reinforced in the minds of the students by folding a strip of paper an unknown number of times using the + fold process. This creates a stack of folds as shown in (a) of Figure 6. Nothing is known about the direction of the folds at either end of the folded stack. It is known that the beginning of the strip containing the mark is at the bottom of the stack. Without knowing the direction of

the folds, it is impossible to code the sequence of turns. To illustrate this fact, the coding could be written as a sequence of question marks.

? ? ? ? ? ... ?

Fold the stack one more time in the manner illustrated in (b) of Figure 6.

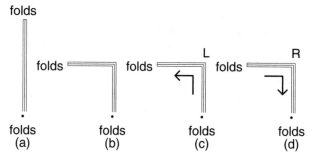

Figure 6: Adding one more fold and noting the directions of the folds.

The folds at either end remain unchanged and unknown (indicated by the plain text question marks), but the new fold creates another sequence of folds (indicated by the bold text question marks) and provides a reference point from which a coding sequence can begin. As shown in (c) of Figure 6, travel along the stack of paper from the beginning point, labeling the direction of the turn (*L*).

L ? ? ? ? ? ... ? ?

At the end of the stack are a number of folds in which the code is again unknown (indicated by the plain text question marks in the coding sequence). By retracing the route (d), the direction of the new fold encountered is *R* and can be coded into the sequence. The new coding is

L ? R ? ? ? ? ... ? ?

At the beginning of the stack, the direction of the folds is unknown again (indicated by the plain text question mark in the sequence). Repeat this process of traveling from one end of the folds to the other several times, as illustrated in Figure 7.

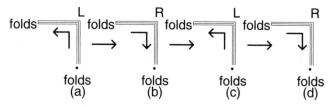

Figure 7: Continuing to unpack the directions of the new folds.

This process results in this coding sequence:

L ? R ? L ? R ? L ? R ? L ? ... ? R

The last turn will always be coded with an **R** because once the beginning of the strip is reached there are no more folds. This activity reinforces the pattern of inserting L and R into the sequence of the code—what is known as the Inflation Law.

Determining whether the crease at a particular position in the paper folding curve has a directional code of L or R can easily be accomplished in the early steps of the paper folding process simply by folding the paper and traveling along the curve. However, when folding the paper becomes impossible or when writing the coding sequence becomes impractical, the directional code of L or R can be determined by using a simple algorithm.

Ask the students the following question: "What is the direction of the twelfth crease?" The code for Step 4 must be written before the twelfth crease can be examined and is found to be an **R**.

L	L	R	L	L	R	R	L	L	L	R	**R**	L	R	R
1	2	3	4	5	6	7	8	9	10	11	**12**	13	14	15

Now ask the question, "What is the direction of the 51st crease?" At this point it should be obvious that writing the coding sequence and counting is not the most efficient way of answering the question. There must be a simpler way to find the answer.

The Inflation Law provides a way of determining the direction of a crease without writing the coding sequence. The first time the L and R are inserted into Step 1, the L is in the first position and the R is in the third position. Position 1 is always an **L**. Position 3 is always an **R** as illustrated below.

$$L$$
$$1$$
$$\downarrow$$
$$\begin{array}{ccc} L & L & R \\ 1 & 2 & 3 \end{array}$$

If the Inflation Law is applied to Step 2, the inserted Ls and Rs appear *in the odd numbered positions* of Step 3 and the previous code fills in the even numbered positions. In fact, the position numbers of Step 2 have just been doubled in Step 3.

Originally, every even numbered position first appeared in a sequence as an odd numbered position. The odd numbered position was then doubled in each subsequent step. Working backward, position 12 can be traced back to position 6

which can be traced back to position 3. That is, position 12 originated as position 3. Therefore, position 12 must be an R because position 3 was an R.

What happens, if in working backwards to the originating position, the resulting position number is not a 1 or a 3? Suppose the position number that is questioned is 68. Position 68 can be traced back to position 34, which can be traced back to position 17. But what code is position 17? There are only fifteen coding symbols in Step 4 so the code of position 17 is still unknown. By concentrating only on the **L**s that were inserted by the Inflation Law, a pattern might be discovered.

L	L	R	L	**L**	R	R	L	**L**	L	R	R	**L**	R	R
1	2	3	4	**5**	6	7	8	**9**	10	11	12	**13**	14	15

The difference between the position number of the **L**s is four. Upon closer inspection, each of the position numbers is one more than a multiple of four.

It appears that whenever an odd number is divided by four and has a remainder of 1, the position will code as an L. According to this hypothesis, position 68 will be an L (17 divided by 4 has a remainder of 1). At this point, it seems safe to say that to find the code for an even numbered position, repeatedly divide the number by two until the quotient is an odd number. Divide the odd number by four and if the remainder is 1, the code of the position is L. What if the remainder is not 1? To answer that question, one must concentrate on the Rs that were inserted by the Inflation Law.

L	L	**R**	L	L	R	**R**	L	L	L	**R**	R	L	R	**R**
1	2	**3**	4	5	6	**7**	8	9	10	**11**	12	13	14	**15**

Again, the difference between the position numbers is four. This time, each of the position numbers is three more than a multiple of four.

Whenever an odd number is divided by four and has a remainder of 3, the position will code as an R. Now, combine these two observations into one rule that applies for odd numbered creases. If the position of the crease is an odd number, divide the number by 4. If the remainder is 1, the code for the crease is L. If the remainder is 3, the code for the crease is R. The students may ask, "What happens if the remainder is 0 or 2—the other possible remainders when dividing

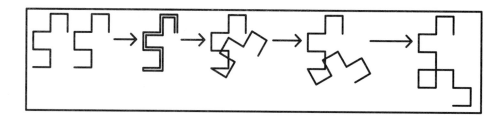

Figure 8: Making a Step 4 curve from two Step 3 curves.

by four?" The remainder could only be a 0 or 2 if the original number that was being divided by four was an even number. The rule stated above always begins with an odd numbered crease. Therefore, a rule for even numbered creases is also necessary: *If the position of the crease is an even number, divide the number by two until the quotient is an odd number and then apply the odd number rule.*

The question that originated this discussion was, "What is the direction of the 51st crease?" Finding the answer should prove easy at this point. Fifty-one is an odd number and the rule for odd numbers is to divide by 4 and look at the remainder. The remainder is 3. Therefore, the direction of the crease is *R*. The rules discovered apply no matter what number crease is being examined. (Since this task is repetitive, using a calculator will be beneficial to students.)

Special notes

The activities in this unit are a combination of two units of the Paper Folding CATE. Being able to determine the code of a particular crease is dependent on understanding the Inflation Law. Therefore, it is suggested that the Inflation Law be introduced and thoroughly examined before the student proceeds to find the code of a particular crease.

2.3   Unit 3. The Dragon curve

Creating a Dragon curve is a hands-on activity that enables students to see rotational symmetry and the self-similarity inherent in fractals. The introduction of the Reflection Law of paper folding enables students to see the symmetry of the coding sequence. This activity works best with cooperative groups of four students.

Grade level: 5–12
Materials: 3 strips of paper (1″ × 11″) per student
                  Transparencies (cut into quarters), 32 pieces
                       are required to complete a Step 9 curve
                  Transparency markers in four colors
                  Transparency tape
                  Transparency of Steps 5–9 (Appendix 7)
                  Overhead projector

Notes to the teacher

Ask the students to fold two **Step 3** curves using two strips of paper. Next, ask the students to arrange the two Step 3 curves to form a Step 4 curve. Allow the students plenty of time to discover how the pieces should be put together. Have the students write a description of the method they used. Did each group use the same method? A volunteer could bring the strips to the overhead and demonstrate the method used. One method to create a Step 4 curve from two Step 3 curves is illustrated in Figure 8.

Begin with two Step 3 curves and nest one curve inside the other. In the nested form, tape the two strips together at the ends pointing west and rotate the inside strip 90° clockwise.

This process is known as the **Assembly Method** for creating curves. The Assembly Method allows for the creation of steps beyond those that are possible by real paper folding. The curve created demonstrates **rotational symmetry**: the figure can be separated into two parts at a point so that when the parts are rotated in some manner, they can be seen as identical. The point of rotation in this case is the aligned end points of the two Step 3 curves. It should also be noted that the Step 4 curve created is not the same size as a Step 4 curve produced by paper folding, but the shape is the same. When figures have the same shape, but not necessarily the same size, they are said to be **similar**.

To examine what happens to the code of the curve when two steps are used to create a new step, it might be easier to decompose a step and look at the parts. Have the students fold a Step 2 curve. Find the midpoint fold and tear the strip of paper apart at this fold (Figure 9 top).

Examine the two pieces that are formed. Each piece is a Step 1 curve, however smaller, but the code of each piece is different. The first part of the curve is read by starting north and ending west, but the second part of the curve is read starting south and ending west (Figure 9 bottom).

One way to explain what is happening to the code of the individual pieces is to relate the code to a mirror. If the midpoint fold serves as the mirror, the *L* that is one letter in front of the mirror reflects to an *R* one letter behind the mirror.

Midpoint fold: **L**

Step 2:   L   **L**   R

Midpoint fold

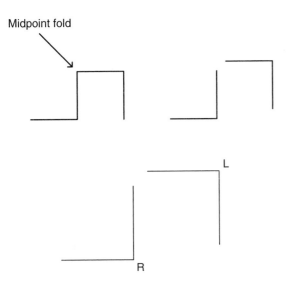

Figure 9: Top: Decomposing a Step 2 curve into two Step 1 curves. Bottom: Labeling the Step 1 curves.

The same thing happens in Step 3. The *R* that is one letter in front of the mirror reflects to an *L* that is one letter in back of the mirror. The *L* that is two letters in front of the mirror reflects to an *R* that is two letters behind the mirror, and so on.

Step 3:   LLR   **L**   LRR

Step 4:   LLRLLRR   **L**   LLRRLRR

The process of replacing an *L* with an *R* and an *R* with an *L* across the midpoint fold is known as the **Reflection Law** for paper folding. The mathematical properties of similarity and symmetry are incorporated in the Reflection Law. The previous step is always in front of the mirror (the midpoint fold) and the use of the Reflection Law accounts for the different coding of the piece behind the mirror. Now, there are two laws—the Inflation Law and the Reflection Law—that can be used to discover the coding for paper folding. It is an interesting challenge to rigorously prove how to go from one law to the other just using mathematical reasoning.

When using the Assembly Method of creating new steps of a paper folding curve, it should be noted that the size of the curve quickly becomes unmanageable and it is difficult to maintain the right angle of the creases without mounting the strips on a board. Allow the students to discover this idea for themselves by having them work in groups to combine two of the assembled Step 4 curves to create a Step 5 curve and use two of the assembled Step 5 curves to create a Step 6 curve. It becomes obvious to the students that it is very difficult to maintain the right angles when assembling the Steps.

One way to create the curves without losing the formation of right angles is to trace one step on a piece of transparency and assemble two pieces together. Break the students into eight groups of four, and give the same color transparency marker to all students in a group. Instruct the students to place the piece of transparency on the copy of the Step 4 curve and *carefully* trace the curve on the transparency. Working in pairs, students should place one transparency piece on top of another, aligning the curves. Keeping the west pointing ends of the steps together, rotate the top transparency 90° clockwise. Tape the two pieces together. See Figure 10.

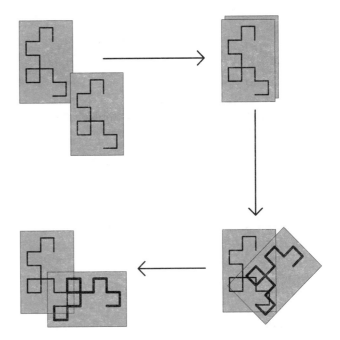

Figure 10: Assembling a Step 5 curve from two Step 4 curves.

Each group of four students now has two Step 5 curves that can be assembled together in the same manner to create a Step 6 curve. Continue to pair the steps in the same manner, keeping the same color for both curves of the pair for as long as possible. In a classroom of 32 students, the eight groups will be able to assemble a curve through Step 9. At this stage of construction, the curve appears to be taking on the shape of a Dragon. Illustrations of the four color curve at Step 6 through Step 9 appear in Figure 11.

This project was created by beginning with a Step 4 curve. If the resources are available, making transparencies of a Step 7 curve for each student will allow a class of thirty-two students to complete the Dragon curve through Step 12.

The use of transparencies solves the problem of the keeping right angles in each stage of construction. What happens to the size of the Dragon presents another problem. Using the Assembly Method, eventually the Dragon will become too large to fit on the desktops or even the wall if several classes are involved in the project. So how are graphics created to show the Dragon curve and still remain on a single page? In real paper folding, the length of each segment created by folding the strip of paper is reduced by a factor of 0.50 from one

Figure 11: Steps 6 through 9 of the Dragon curve, in four colors. See the color plates.

step to the next. The paper strip itself remains the same length but the space that the curve occupies is reduced in each subsequent step as the beginning point and endpoint become closer together. In the Assembly Method, the length of each segment remains the same but the length of the curve is increased by a factor of 2 from one step to the next, and the space occupied by the curve is rapidly increasing as the beginning point and endpoint separate. In order to avoid shrinking to a point (the limit in real paper folding) or expanding to infinity (the limit in the Assembly Method), another factor must be used that will adjust the length of each segment in the curve while maintaining the distance between the beginning point and the endpoint.

If the strip of paper (Step 0) in the paper folding process is thought of as a rubber band of length a, then Step 1 can be created by "pulling" the rubber band to form a right angle with sides of length b. See Figure 12.

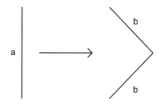

Figure 12: An elastic version of Step 1.

From the Pythagorean theorem we see the lengths of a and b are related by $b = a/\sqrt{2}$.

The computer generates the Dragon curve by using a scaling factor of $1/\sqrt{2}$. The distance between the beginning point and endpoint remains constant as the length of the segments is being scaled. In Figure 13, Steps 1 through 5 illustrate this fact as well as showing how the curve is being rotated 45° from one step to the next, not the 90° as in real paper folding. The computer is a necessary component of *scaled paper folding*.

The hands-on activity of creating a Dragon curve can be used to introduce the concept of *self-similarity*. The figure is created from two identical parts. Each of these parts is the previous step. As illustrated in Figure 14, in the beginning

steps of the formation of the curve, the two identical parts are obviously different from the entire curve.

In later steps in the formation of the curve, the differences between the two parts and the entire curve become smaller and as illustrated in Figure 15, the parts *appear* to be similar to the entire curve.

If the red (lighter) and blue (darker) parts from the curves above are separated into the steps from which they were created, curves of four colors are generated. These smaller steps are identical to each other and appear to be similar to the entire curve. See Figure 16.

Continuing only one step further, the differences in the parts and the entire curve are becoming so small that the parts appear to be a smaller copy of the entire curve, but in reality they are still not similar. (The step 10 curve shown in Figure 17 has been magnified for illustration purposes.)

It is not until the figure reaches its limit (Figure 18) that the parts are truly similar to the entire curve. When the parts of an object are similar to the whole object, the object illustrates *self-similarity*. If one of the smaller parts is magnified to the same size as the entire curve, the two would look identical!

The activities in this unit cover a variety of mathematical concepts that include similarity, reflections, rotational transformations, and symmetry. In constructing the Dragon curve, students can begin with any step of the paper folding process, duplicate the step, and create the next step by using rotational symmetry. If the creation process is continued to further and further steps, the curve begins to demonstrate the property of self-similarity.

3   Fractal wallhanging

Self-similarity and symmetry are the underlying notions for the Fractal Wallhanging CATE. Students have more than likely encountered symmetry, but when referring to fractals, symmetry is approached from a slightly different perspective. The idea of self-similarity may be new to most students.

A physical object or figure is said to have symmetry if it can be broken into parts that are the same. It may be necessary to accept that the parts are loosely the same. The parts may not be identical immediately; that is, one part might need to

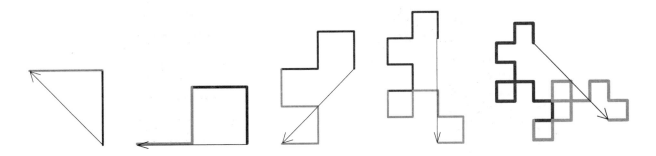

Figure 13: Successive rotations in Steps 1 through 5 of the scaled paper folding approach to generating the Dragon curve. See the color plates.

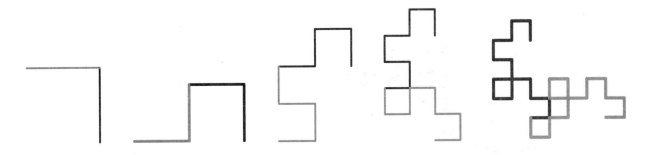

Figure 14: Decomposing the early stages of the Dragon curve into two parts. See the color plates.

Figure 15: Steps 6 through 9 of the Dragon curve, in two colors, illustrating the self-similarity of the curve. See the color plates.

Figure 16: Steps 6 through 9 of the Dragon curve, in four colors. See the color plates.

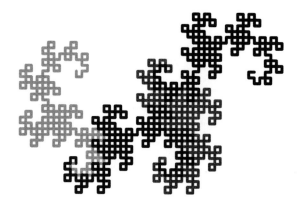

Figure 17: Step 10 of the Dragon curve, in four colors. See the color plates.

Figure 18: The Dragon curve, limiting shape. See the color plates.

Translational Symmetry

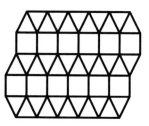

Reflectional Symmetry

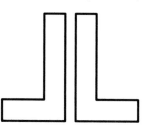

Rotational Symmetry

Figure 19: Three familiar symmetries.

be moved (rotated, reflected, translated) in order for it to be in the identical orientation as the other part. The *symmetry* is the movement required to make the parts look the same. See Figure 19.

Fractals take the notion of symmetry one step further: self-similarity. When a fractal is broken into parts, in addition to the parts being identical, each part looks like the whole. A demonstration with a head of cauliflower will help make this more clear. A head of cauliflower is not self-similar in the pure, mathematical sense, yet it clearly demonstrates the property without needing mathematics to explain it. The cauliflower head contains branches that, when broken off the head, look very much the same as the whole head, only smaller. This broken off branch contains smaller heads that also look very similar to the first branch, as well as to the whole head of cauliflower. This process of breaking off heads can be done only about three or four times before the pieces get too small to break further. Demonstrating the first few stages is an ideal way to give students an intuitive feel for self-similarity. It may also be pointed out that when dealing with mathematical fractals, self-similarity occurs at in-

finitely many scales (even microscopic), unlike real, physical objects.

In Unit 1 of Fractal Wallhanging, the activities focus on analyzing and decoding mathematical fractal images. The notion of self-similarity is intuitively introduced as the students discover the code and recreate fractal images. Students will discover the images can be decomposed into parts that are exact replicas of the whole. Students often are fascinated by the idea that images can be created by simply putting together three smaller copies of themselves in a certain manner.

In Unit 2, students are actively involved in constructing two large fractal images (up to 5 feet square). The power of iteration is revealed to students as they repeatedly put three images together, according to the code of the image, to create the next stage of the fractal. With the creation of the second wallhanging, students should clearly understand that the code of a fractal is the determining factor in its appearance. The concept of self-similarity is clearly displayed as students build successive stages of the fractal. If *ClarisWorks* or some similar draw program is available, students can build a fractal image by following the iterative process of Reduce, Repli-

Figures 11 and 16, Chapter 9: Steps 6 through 9 of the Dragon curve, in four colors.

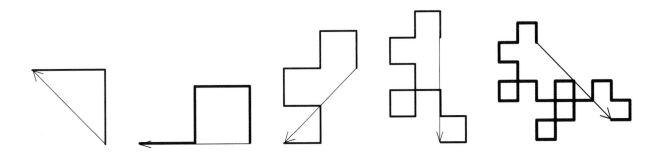

Figure 13, Chapter 9: Successive rotations in Steps 1 through 5 of the scaled paper folding approach to generating the dragon curve.

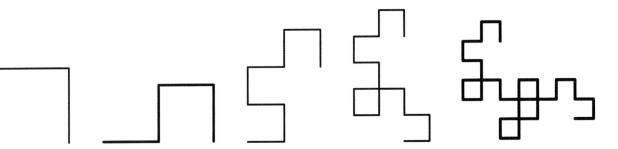

Figure 14, Chapter 9: Decomposing the early stages of the Dragon curve into two parts.

Figure 15, Chapter 9: Steps 6 through 9 of the Dragon curve, in two colors, illustrating the self-similarity of the curve.

Figure 17, Chapter 9: Step 10 of the Dragon curve, in four colors.

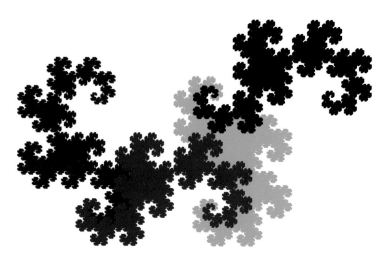

Figure 18, Chapter 9: The Dragon curve, limiting shape.

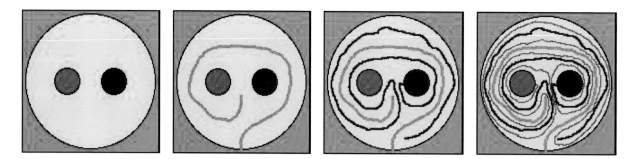

Figure 1, Chapter 6: The first three branches of the Wada canals. After infinitely many have been dug, every point of the island is on the edge of a canal to the ocean and to each lake.

Figure 5, Chapter 6: Left: the pattern revealed when blue and red light shines through two openings. Right: all three lights illuminating the interior of the region.

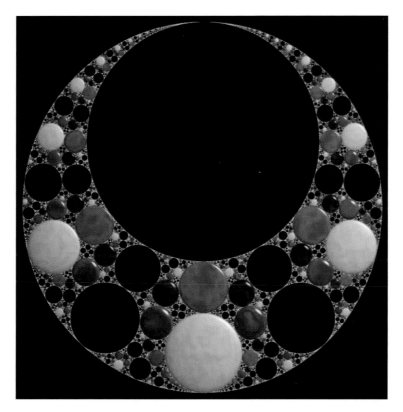

Figure 4, Chapter 2: Pharaoh's breastplate.

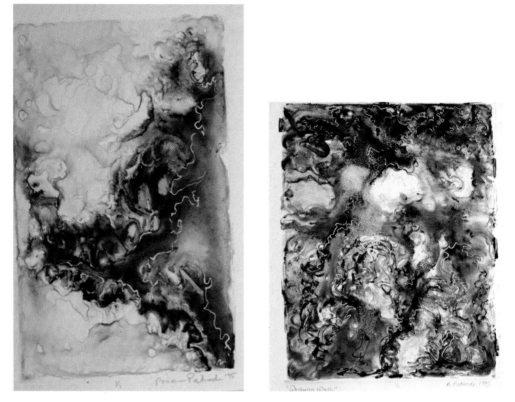

Figure 8, Chapter 14: Fractal monoprints. Left, "Untitled"; right, "Wishing well."

Figure 9, Chapter 14: Fractal collagraphs, both "Untitled."

cate, Rebuild. Appendix 8 is *Chaos*, a program for the *TI-82* or *TI-83* graphing calculator. This program allows students to create fractal images by selecting the transformations (the code) to be applied in each cell of the image. The program is called Chaos because it uses the Chaos Game principle to plot the points.

The fractal images analyzed in these activities are generated by reducing images to half-size, making three copies of the reduced image, and putting the images together by following a certain code. The Multiple Reduction Copy Machine (Peitgen, Jurgens & Saupe (1992a)) metaphor can be used to explain this process. The copy machine takes an image as input. It has several independent lens systems, each of which reduces the input image and places it somewhere in the output image. The assembly of all reduced copies in some pattern (the code) is finally produced as output. The critical idea is that the machine runs in a feedback loop; its own output is fed back as its new input again and again. It is important to understand that no matter what initial image is taken, when put through the iterative process of the Multiple Reduction Copy Machine, the sequence of images obtained always tends toward the same final image.

3.1 Unit 1. Reduce, replicate, rebuild

Activity 1: Solve the Puzzle 1

In this activity, students are introduced to fractal images and are given the opportunity to experiment with fractal puzzles. Students begin analyzing fractals in a concrete and non-mathematical manner. Students are given fractal images and are asked to solve the puzzle—that is, determine what must be done to three smaller copies of the image to recreate the image.

Grade level: 5–12
Materials: One copy of **Solve the Puzzle 1** activity sheet
         per student (Appendix 9)
         Scissors
         Glue, glue stick or transparent tape
         Overhead transparency of the **Solve the
         Puzzle 1** activity sheet for classroom
         demonstration
         Overhead transparency of the **Image Without
         Grid** (Appendix 10)

Notes to the teacher

In the **Solve the Puzzle 1** activity, students study a fractal image and place the three pieces into the grid to make a duplicate of the image. Students are asked to describe, in simple language, how they manipulated the pieces to recreate the image.

Students begin by cutting out the three reduced images at the bottom of the activity sheet. They may notice these are reduced size copies of the large image at the top of the sheet. Next, students place the three images in the empty grid so that when completed, the image created is identical to the large image. Students should write their description of how they solved the puzzle before they glue the pieces in place. As in a jigsaw puzzle, the pieces may need to be turned to make them fit.

The concept, that an image can be recreated by putting together three smaller copies of itself, in some manner, is incredible!! Self-similarity, is an important notion that is easily overlooked. Many objects or figures are made up of copies of themselves. Some examples that occur in nature are cauliflower, broccoli, clouds, mountain ranges, and fern leaves. For natural objects, the precise and exact notion of self-similarity is somewhat relaxed, but it is still a remarkable and important observation.

In the example shown in Figure 20, square D is empty and the grid is not part of the image; it provides a framework to order the image. Students tend to view the original image as three subparts, not as a single entity. Display a transparency of the image without the grid to help make this clear to students.

Activity 2: Solve the Puzzle 2

In **Solve the Puzzle 1**, students are given fractal images and are asked to solve the puzzle. The same basic principles apply to this activity except in **Solve the Puzzle 2**, unlike a jigsaw puzzle, in addition to rotating the smaller pieces, students are allowed to flip over the three small pieces to replicate the larger image at the top of the page. In addition, students begin to shorten their sentence descriptions into a mathematical shorthand.

Grade level: 5–12
Materials: One copy of **Solve the Puzzle 2** activity sheet
         per student (Appendix 11)
         Scissors
         Glue, glue stick or transparent tape
         Overhead transparency of the **Solve the
         Puzzle 2** activity sheet for classroom
         demonstration
         Overhead transparency of the **Image Without
         Grid** (Appendix 10)

Notes to the teacher

In **Solve the Puzzle 2**, students begin by cutting out the three reduced copies of the larger image, called the tool, at the bottom of the sheet. The tool has a front and a back side. This design allows more options when solving the puzzle. The letter *I* (Identity) indicates the smaller copy is identical to the large image when the *I* tab is pointed down. The letter *V*

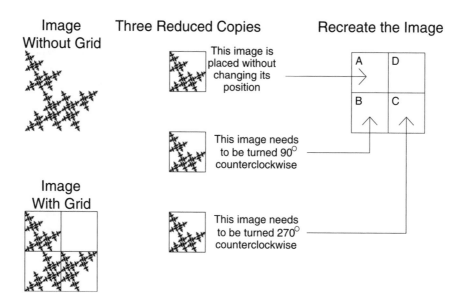

Figure 20: Solve the Puzzle 1: building an image from smaller copies of itself.

(Vertical) indicates that the small copy is the reflection across its vertical axis of the large image when the *V* tab is pointed down. See Figure 21. Note *I* and *V* are two of the eight symmetry transformations of the square. The other six will be introduced in the next activity. Symmetry transformations can be described as movements or actions that can be applied to an image in order to change its orientation.

Once the tools are carefully cut out, students fold each tool along the dotted line so there is an image on the front and back. Glue or tape the sides together so that the tools will stay closed during the manipulation process. Once the three tools are assembled, students will discover (as in **Solve the Puzzle 1**) what to do with each tool to create an image identical to that at the top of the page. In the middle of the activity sheet, the students give a written description of how they manipulated each of the three small copies to replicate the image. In the left-hand column, students are to record, or describe, how they have manipulated each tool to get proper placement in each cell for a successful copy. In the right-hand column, students should write a more mathematical description to describe the steps they followed. These descriptions provide a form of mathematical shorthand.

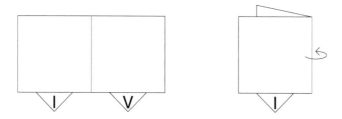

Figure 21: The tool for reflection across the vertical axis.

Activity 3: Find the Code

In this activity, students will extend the idea of **Solve the Puzzle 2** to include mathematical notation of all eight of the symmetry transformations as they "Find the Code" of the fractal image.

Grade level: 5–12

Materials: One copy of **Blank Tool** per student
                          (Appendix 12)
                          One copy of each **Find the Code** activity
                          sheet per student (Appendix 13)
                          Scissors
                          Glue, glue stick or transparent tape
                          Overhead transparency of the **Find the Code**
                          activity sheet for classroom demonstration
                          Overhead transparency of the **Image Without
                          Grid** (Appendix 10)

Notes to the teacher

The tool used to find the code of the fractal image is shown on the left of Figure 22. Students will fill in the tabs and the tool will then look like the picture on the right of Figure 22.

Shown in Figure 23, the eight symmetry transformations are

I: Identity No rotation or reflection. The image on the tool is identical to the large image; the **I** tab is facing front and pointing down. This is the Identity position.

In all remaining cases, begin with the tool in the Identity position.

R_{90}: Rotation 90 degrees counterclockwise Rotate the tool 90° counterclockwise. Label the tab pointed down **R_{90}**.

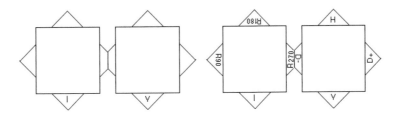

Figure 22: Making the complete tool, with all eight symmetries.

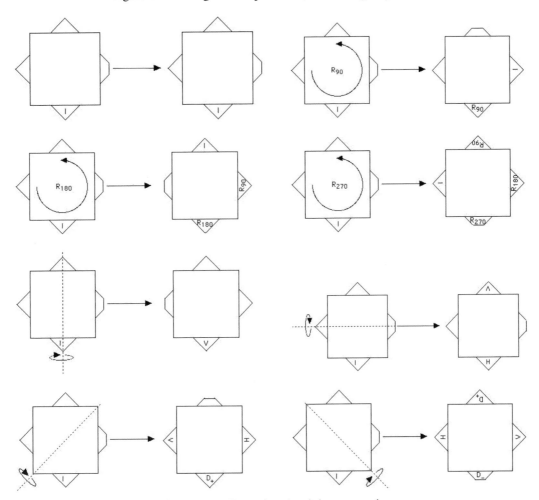

Figure 23: Illustrating the eight symmetries.

R₁₈₀: Rotation 180 degrees counterclockwise Rotate the tool 180° counterclockwise. Label the tab pointed down **R₁₈₀**.

R₂₇₀: Rotation 270 degrees counterclockwise Rotate the tool 270° counterclockwise. Label the tab pointed down **R₂₇₀**.

V: Vertical Reflection Reflect the tool over the vertical axis. The **V** tab will be facing front and pointed down.

H: Horizontal Reflection Reflect the tool over the horizontal axis. Label the tab pointed down **H**.

D_+: Positive Diagonal Reflection Reflect the tool over the positive slope diagonal axis. Label the tab pointed down D_+.

D_-: Negative Diagonal Reflection Reflect the tool over the negative slope diagonal axis. Label the tab pointed down D_-.

In **Solve the Puzzle 2**, students described how they manipulated the tools in terms of I and V to make some kind of turn or flip. With the development of the six other transformations, students can now begin using mathematical notation to

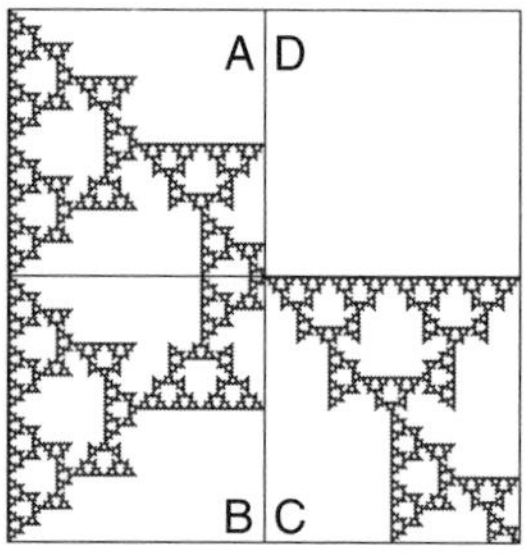

	3-MOVES	2-MOVES	1-MOVE
CELL A:	I, H, V, R_{180}	I, R_{90}, R_{270}	I
CELL B:	I, R_{270}, V, R_{90}	I, R_{90}, D_+	H
CELL C:	I, R_{90}, V, R_{180}	I, H, R_{270}	D.
CELL D:	--- EMPTY ---		

Figure 24: Solve the Puzzle 2, an example.

describe the moves needed to solve the puzzle. Remind them to begin all moves and statements with the Identity position.

The activity sheet is designed with three columns headed: 3 moves, 2 moves, 1 move, to allow students the opportunity to write down multiple solutions for each cell. Figure 24 shows an example.

A discussion with the students about why so many solutions are possible is important. Students may not understand that the puzzle can always be solved with just one move per cell. The reason for this is that any composition of two transformations from I, R_{90}, R_{180}, R_{270}, V, H, D_+, D_- yields one of these transformations again. Therefore, if the puzzle is solved with several moves, students can work the solution down to just one move step by step.

In conclusion, upon completing the activities in Unit 1, it is important that students have made connections and understand the following points:

- the fractal images are a single unit,

- the fractal images can be broken into three smaller parts, each part is identical to the original image, only smaller (each small part is similar to the original image),

- in order to recreate the original image from the three smaller images, it might be necessary to apply one of the eight symmetry transformations to one or all of the three smaller images.

3.2 Unit 2. Building the fractal wallhanging

Activity 1: Fractal wallhanging from limit images

In Unit 1, students solved fractal puzzles by determining the transformations that were applied to the reduced copies of the original image to recreate the image. As students begin Unit 2, they should be familiar with the idea of self-similarity. The activity in this lesson brings to life the process of generating fractal images through iteration. In Unit 1, once students determined the transformations needed to recreate the image, their task was done. In the creation of a fractal wallhanging, determining the transformations is just the beginning. Once

the transformation, or code, is determined for each cell, students repeat the process of putting the pieces together to form successive stages of the image as described below. One other difference in this activity is that when constructing the fractal wallhanging, the image is not reduced before the three copies are made. Because there is no reduction, the image at each successive stage will be twice as large as the image of the previous stage.

Grade level: 5–12

Materials: Overhead transparency of the **Image Without Grid** (Appendix 10)

        Copies of the **Wallhanging "Limit" Images** sheets (Appendix 14)

        Copies of the **Grid Sheet** (Appendix 15)

        Blank paper or chart paper

        Scissors

        Glue stick or glue

        Transparent tape

Notes to the teacher

Building the Fractal Wallhanging is almost like making a quilt. The pieces of the quilt are put together according to a predetermined pattern, in this case the code of the fractal image.

Step 1—Determine the code

Begin with a fractal image. The transformation, or code, for each cell of this image is shown in Figure 25.

Step 2—Construct Stage 1

Begin constructing the wallhanging. Three copies of the image and one Grid Sheet square are needed.
Glue or tape the pieces onto the Grid Sheet. At this point a **Stage 1** image has been completed. See Figure 26.

Step 3—Construct Stage 2

Repeat the process with three Stage 1 images.
Tape the pieces together. At this point a **Stage 2** image has been completed. See Figure 27.

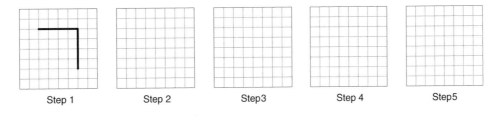

Figure 25: A fractal image, with the symmetries of the pieces indicated.

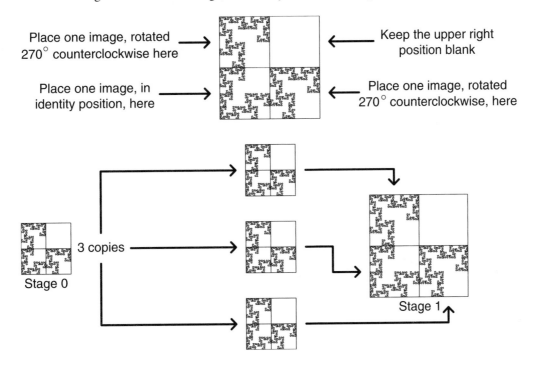

Figure 26: Stage 1 of the fractal wallhanging.

Step 4—Continue construction process

Stages 3 and 4 can be completed in the same manner. Typically, a fractal wallhanging can be constructed to Stage 5, 6, or higher. The size of the final stage will depend on the size of Stage 0, but usually will be 5 to 6 feet square.

Preparation:

1. Decide the size of the fractal wallhanging to be constructed. Use this information to determine to what stage the fractal image will be built.

2. Determine the number of Grid Sheet (Stage 1) squares each student will need for the construction. Generally, a class can easily build an image to Stage 5, which means 81 Stage 1 images will need to be made.

3. Measure and cut the appropriate number of blank squares for each Stage. (81 blanks for Stage 1, 27 blanks for Stage 2, nine blanks for Stage 3, and three blanks for Stage 4.)

Activity 2: Wallhanging from one filled square

In this activity, students will create a Fractal Wallhanging of a fractal image that has the exact same code as the image in Activity 1. This wallhanging, however, will be created from Iteration 5 images that have been generated from iterating the code when starting with one filled square. The students will be building a fractal from scratch.

Grade level: 5–12
Materials: **Progression of a Fractal Image** sheet
   (Appendix 16)
   Copies of the **Wallhanging Iteration 5 Images** sheets (Appendix 17)
   Copies of the **Grid Sheet** (Appendix 15)
   Blank paper and/or chart paper
   Scissors
   Glue stick or glue
   Transparent tape

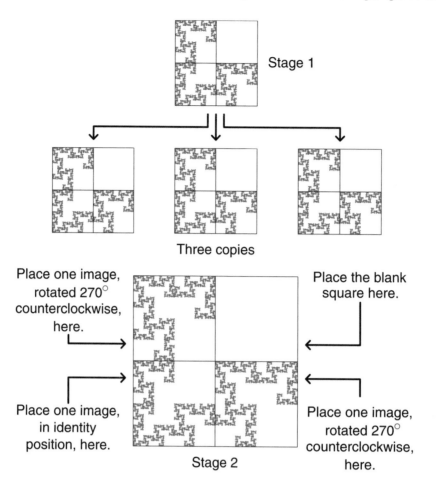

Figure 27: Stage 2 of the fractal wallhanging.

Notes to the teacher

Display the Progression of a Fractal Image sheet (Appendix 16) on the overhead projector. These images show the progressive stages of the fractal beginning with one filled square. This image has the same code as the Fractal Wallhanging Limit image. In the early stages (Iterations 1, 2, 3), the image is very crude and has little detail. But by the time Iteration 5 is reached, the image has enough detail that when compared to the limit image, it is apparent they are derived from the same code.

By continuing to iterate to Iteration 6, 7, 8, etc., more and more detail will emerge and the image looks almost the same as the limit image. When the code is iterated an infinite number of times, the limit image would be reached.

A second wallhanging can be constructed, following the same steps as before, this time using the Iteration 5 images. When the two wallhangings are hung side by side and viewed from a distance, it is remarkable how close they are in appearance.

In constructing the Fractal Wallhanging, students began with one fractal image (approximately 2 inches square) at Stage 0 and created a fractal image 5 to 6 feet square at Stage 5 by iterating the process of putting three images together according to the code of the image. The wallhanging created in *Fractal Wallhanging from Limit Images* was created from what are called limit images. This means the images are the product of iterating the code of the image an infinite number of times. Although in reality iterating an image an infinite number of times is not possible, the limit images are a physical representation of the theoretical infinite iterations. At this point, when comparing the image from one stage to the next, one would not see any differences in the images. The power of iterating the code may be easily overlooked. It is natural that this large image appears since it was created from reduced copies of itself. What is amazing, and unbelievable to many, is that the same large fractal image will also be created by starting with a filled black square, as demonstrated in *Wallhanging from One Filled Square*. The starting image is irrelevant. It is the *code* and the *iterative process* that produces the wonderfully beautiful fractal image!

Appendix 1

# of Folds	# of Segments	
1	$2 = 2$	$= 2^{--}$
2	$4 = 2 \times 2$	$= 2^{--}$
3	$8 = 2 \times 2 \times 2$	$= 2^{--}$
4	$16 = 2 \times 2 \times 2 \times 2$	$= 2^{--}$
5	$32 = 2 \times 2 \times 2 \times 2 \times 2$	$= 2^{--}$

Transparency # 1

Appendix 2

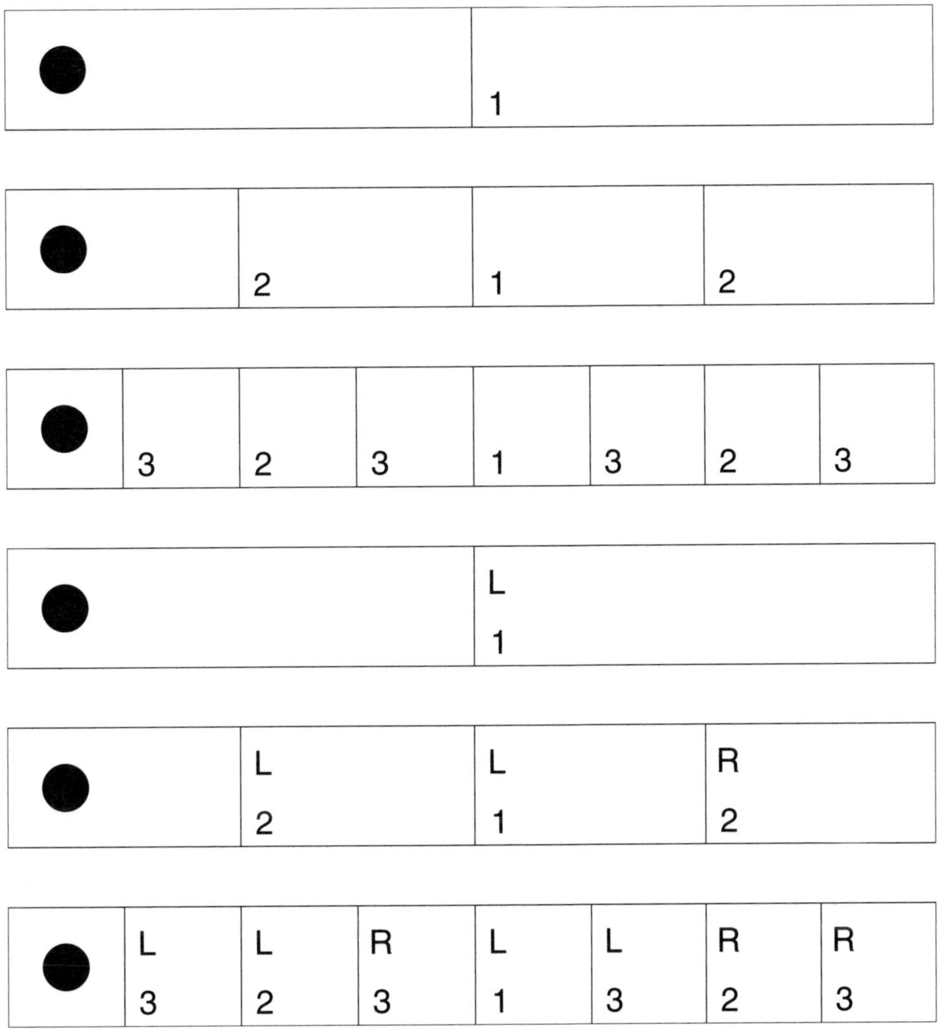

Transparency # 2

Appendix 3

Student Activity Sheet 1

1. Using the paper folding process, complete the chart to folds 1–3.

# Folds	# Segments	# Creases
1		
2		
3		
4		
5		
6		
7		
8		
9		
10		
⋮		
n		

2. Predict the number of segments and creases for fold 4 and write your prediction in the chart. Verify your prediction by folding a strip of paper.

3. Describe in words how you arrived at your prediction for the number of segments.

4. Describe in words how you arrived at your prediction for the number of creases.

5. Why is it not possible to verify your predictions by folding paper for further steps?

6. Fill in the chart for folds 5–10.

7. Find the formula for the number of segments formed by the nth fold in terms of n.

8. Find the forumla for the number of creases formed by the nth fold in terms of n.

9. How many segments and how many creases are formed by the 15th fold? the 25th fold? the 33rd fold?

	# Segments	# Creases
15th fold		
25th fold		
33rd fold		

10. Suppose the number of segments formed is 536,870,912. How many folds were made?

11. Suppose the number of creases formed is 524,287. How many folds were made?

12. If a stack of 500 sheets of paper is approximately two inches high, how high is a strip of paper that has been folded 10 times? 20 times? 50 times?

13. Step 1 of the paper folding process has been drawn on the graph paper. Draw Steps 2–5 on the graph paper provided and record the direction of each crease.

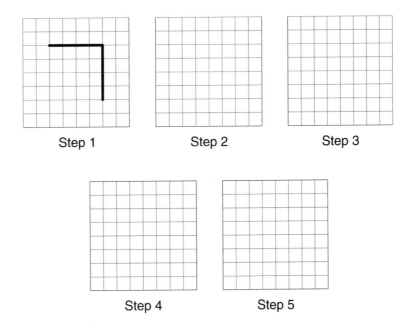

Step 1   Step 2   Step 3

Step 4   Step 5

14. Did you fold paper before drawing Step 4 and 5? Explain why it is or is not possible to draw the figure of each step without actually folding a piece of paper.

15. In folding real strips of paper, how does the length of each segment change from one step to the next?

Appendix 4

Student Activity Sheet 2

1. Define the Inflation Law.

2. Demonstrate that you understand the principle of the Inflation Law by inflating the sequence

LLRLLRRLLLRRLRRLLLRLLRRRLLRRLRR

3. Arrange the following steps in the order necessary to find the direction of a crease:

If the remainder is 1, the crease is L.
If the number is odd, divide by 4.
Determine if the number is odd or even.
Check the remainder.
Apply the rule for odd numbers.
If the number is even, continue to divide by 2 until the quotient is odd.
Select a crease number.
If the remainder is 3, the crease is R.

4. Find the direction of the numbered crease:

      a. 262

      b. 2051

      c. 17,664

      d. 1,048,576

      e. 15,384,209

Appendix 5

Answers to Activity Sheet 1

1, 2, 6.

# Folds	# Segments	# Creases
1	2	1
2	4	3
3	8	7
4	16	15
5	32	31
6	64	63
7	128	127
8	256	255
9	512	511
10	1024	1023
$\vdots$		
n	2^n	$2^n - 1$

3. The number of segments doubles in each step from the previous step.
4. The number of creases is one fewer than the number of segments in the step.
5. Paper can be physically folded only seven times.
7. # Segments $= 2n$, $n =$ # folds
8. # Creases $= 2n - 1$, $n =$ # folds
9.

	# Segments	# Creases
15th fold	32,768	32,767
25th fold	33,554,432	33,554,431
33rd fold	8,589,934,592	8,589,934,592

10. $n = 29$
11. $n = 19$
12. A little more than 4 inches; approx. 350 feet (a twenty story building); approx. 71 million miles

13.

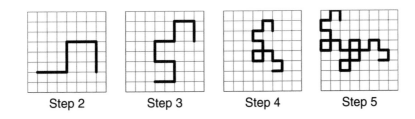

Step 2   Step 3   Step 4   Step 5

    Step 2 LLR
    Step 3 LLRLLRR
    Step 4 LLRLLRRLLLRRLRR
    Step 5 LLRLLRRLLLRRLRRLLLLRLLRRRLLRRLRR

14. At this time, no formal process has been explained that would enable students to draw the figures without first folding the paper, so probably they will say it is necessary to fold the paper.

15. The length of each segment decreases by 0.5.

Appendix 6

Answers to Activity Sheet 2

1. The Inflation Law inserts the LR coding between each symbol of the code in an alternating manner to produce the coding for the next step (LXRXLXRXLXRXLXR ... where X represents the symbols in the previous code).

2. **LLRLLRRLLLRRLRRLLLRLLRRRLLRRLRRLLLRRLRRRLLRLLRRRLLRRLRR**

3. One solution is listed below.

    Select a crease number.
    Determine if the number is odd or even.
    If the number is odd, divide by 4.
    Check the remainder.
    If the remainder is 3, the crease is R.
    If the remainder is 1, the crease is L.
    If the number is even, continue dividing by 2 until the quotient is odd.
    Apply the rule for odd numbers.

4.   a. R
       b. R
       c. L
       d. L
       e. L

Appendix 7

Steps 4, 5, 6, 7, 8, and 9

Appendix 8

PROGRAM: SCALED (TI-83)

```
:ClrDraw:FnOff
:AxesOff:Degree
:PlotsOff
:0- >Xmin:1- > ΔX
:0- >Ymin:1- > ΔY
:0- >R:32- >D:0:- >N
:31- >U:31- >V
:ClrHome
:Disp "TYPE IN THE
RULE"
:Disp "+:LEFT -:RIGHT"
:10- >dim(L2)
:0- >Z:0- >K
:Repeat K=105
:getKey- >K
:If K=95 and Z<10:Then
:Z+1- >Z:1- >L2(Z)
:End
:If K=85 and Z<10:Then
:Z+1- >Z:0- >L2(Z)
:End
:If K=24 and Z>0:Z-1- >Z
:For(I,1,Z)
:If L2(I)=1:Output(3,I,"+")
:If L2(I)=0:Output(3,I,"-")
:End
:Output(3,Z+1">")
:End
:Z- >dim(L2)
:Line(U,V,U+D,V):Pause
:
:Lbl M
:Menu("TYPE MENU:",
"FORWARD",A,
"BACKWARD",B,
"STAGE NO",C, "QUIT",Q)
:
```

```
:Lbl A
:If N≥10:Goto S
:N+1- >N:Goto Z
:Lbl B
:If N≤0: Goto S
:N+1 - >N Goto Z
:
:Lbl C
:Input"STAGE NO:",N
:If N<1 or N>10
:Goto S
:
:Lbl Z
:ClrDraw
:D/2∧(N/2)- >L
:N- >dim(L1)
:dim(L2)- >P
:For(I,1,N)
:I- >H
:While H>P:H-P- >H:End
:L2(H)- >L1(I)
:End
:-(sum(L1)+sum(L1-1))/8
- >R
:While R<0:R+1- >R:End
:While R>1:R-1- >R:End
:31- >U:31- > V
:round(L*cos(360R),0)
- >A
:round(L*sin(360R),0)
:- >B
:Line(U,V,U+A,V+B):U+A
- >U:V+B- >V
:For(I,1,2∧N-1)
:I- >K:0- >A
:While fPart(K/2)=0
:K/2- >K:A+1- >A
```

```
:End
:fPart(K/2)- >C
:If L1(N-A)=0:1-C- >C
:fPart(R+C)- >R
:round(L*cos(360R),0)
- >A
round(L*cos(360R),0)
- >B
:Line(U,V,U+A,V+B):U+A
- >U,V+B- >V
:End
:Text(0,0,N)
:For(I,1,N)
:If L1(I)=1
:Then
:Text(0,4+4I,"+")
:Else
:Text(0,4+4I,"-")
:End
:End
:
:Lbl S
:DispGraph
:Pause:Goto M
:
:Lbl Q
:ClrHome
```

The TI-82 version of SCALED is obtained by replacing the lines : with blank lines.

PROGRAM: CHAOS (TI-83 and TI-82)[1]

```
:ClrDraw
:PlotsOff
:FnOff
:AxesOff
:Lbl 7
:{1,0,0,1}- >L1
:{1,0,0,1}- >L2
:{1,0,0,1}- >L3
:Lbl 6
:ClrDraw
:Disp "CHOOSE A REGION"
:Disp ""
:Disp "1.UPPER LEFT"
:Disp "2.LOWER LEFT"
:Disp "3.LOWER RIGHT"
:Disp "4.DRAW"
:Disp "5.QUIT"
:Input S
:If S=4
:Then
:getKey=0
:Goto 0
:End
:If S=5
:Stop
:Disp "1.IDENTITY"
:Disp "2.ROTATE 90"
:Disp "3.ROTATE 180"
:Disp "4.ROTATE 270"
:Disp "5.HORIZONTAL"
:Disp "6.VERTICAL"
:Disp "7.DIAGONAL 1"
:Input "8.DIAGONAL 2 ?",K
:If K=1:Goto H
:If K=2:Goto A
:If K=3:Goto B
:If K=4:Goto C
:If K=5:Goto D
:If K=6:Goto E
:If K=7:Goto F
:If K=8:Goto G
:Lbl H
:{0,1,1,0}- >L4
:Goto 5
:Lbl A
:{0,-1,1,0}- >L4
```

```
:Goto 5
:Lbl B
:{-1,0,0,-1}- >L4
:Goto 5
:Lbl C
:{0,1,-1,0}- >L4
:Goto 5
:Lbl D
:{1,0,0,-1}- >L4
:Goto 5
:Lbl E
:{-1,0,0,1}- >L4
:Goto 5
:Lbl F
:{0,1,1,0}- >L4
:Goto 5
:Lbl G
:{0,-1,-1,0}- >L4
:Goto 5
:Lbl 5
:If S=1
:L4- >L1
:If S=2
:L4- >L2
:If S=3
:L4- >L3
:Goto 6
:Lbl 0
:-.75- >Xmin
:.75- >Xmax
:-.5- >Ymin
:.5- >Ymax
:rand- >P
:rand- >Q
:Lbl 1
:If getKey=105
:Goto 9
:P- >X
:Q- >Y
:Pt-On(X,Y)
:rand- >N
:If N<.3333
:Goto 2
:If N>.6666
:Goto 4
:Goto 3
```

```
:Lbl 2
:L1(1)X+L1(2)Y- >P
:L1(3)X+L1(4)Y- >Q
:.5P-.25- >P
:.5Q+.25- >Q
:Goto 1
:Lbl 3
:L2(1)X+L2(2)Y- >P
:L2(3)X+L2(4)Y- >Q
:.5P-.25- >P
:.5Q-.25-¿Q
:Goto 1
:Lbl 4
:L3(1)X+L3(2)Y- >P
:L3(3)X+L3(4)Y- >Q
:.5P+.25- >P
:.5Q-.25- >Q
:Goto 1
:Lbl 9
:Line(-.5,0,.5,0)
:Vertical 0
:Pause
:Vertical .25
:Vertical -.25
:Line(-.5,.25,.5,.25)
:Line(-.5,-.25,.5,-.25)
:Line(-.5,.5,.5,.5)
:Line(-.5,-.5,.5,-.5)
:Vertical .5
:Vertical -.5
:Pause
:ClrHome
:Disp "1.CONTINUE"
:Disp "2.MAIN MENU"
:Disp "3.QUIT"
:Input W
:If W=1
:Then
:getKey=0
:Goto 0
:End
:If W=2:Goto 7
:If W=3:Stop
```

---

[1] This program generates images by the random IFS algorithm.

Appendix 9

SOLVE THE PUZZLE 1

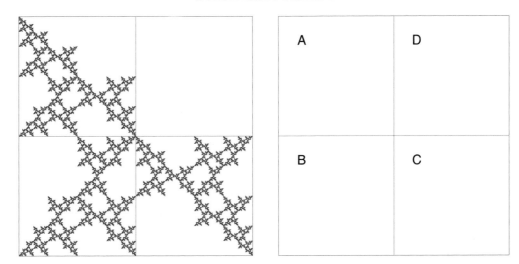

1) SOLVE IT: Cut out the three images at the bottom of the sheet and place them in the grid in order to replicate the image on the left.

2) WRITE ABOUT IT: Write one or two sentences that describe how each reduced image was placed in the appropriate cell.

3) Glue or tape the pieces in place.

CELL A:

CELL B:

CELL C:

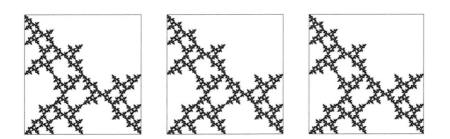

Appendix 10

IMAGES WITHOUT GRIDS

SOLVE THE PUZZLE 1

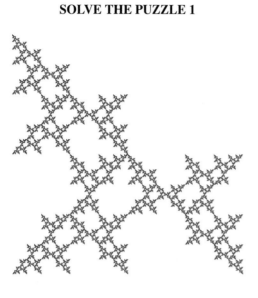

SOLVE THE PUZZLE 2

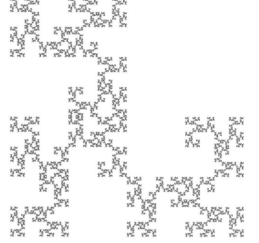

FIND THE CODE

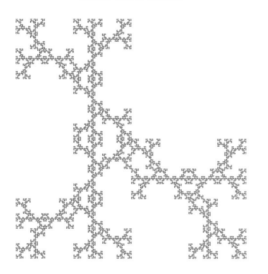

FRACTAL WALLHANGING

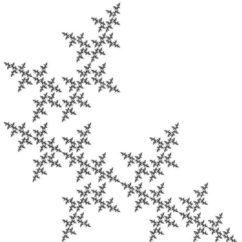

IMAGES WITH CODES

SOLVE THE PUZZLE 1

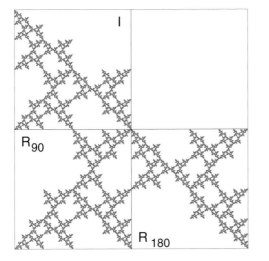

SOLVE THE PUZZLE 2

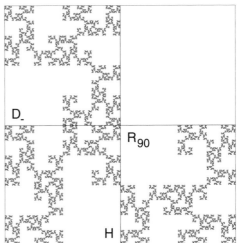

FIND THE CODE

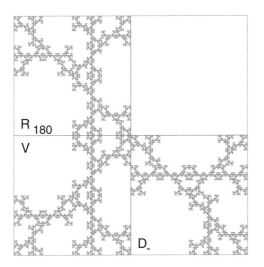

FRACTAL WALLHANGING

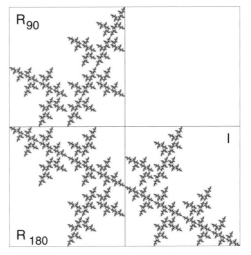

Appendix 11

SOLVE THE PUZZLE 2

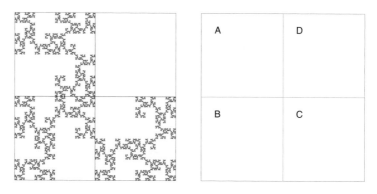

1) SOLVE IT: Cut out the three tools at the bottom of the sheet. Fold each tool along the dotted line and glue or tape the sides together. Place them in the grid in order to replicate the image on the left.

2) WRITE ABOUT IT: In the left hand column, write one or two sentences that describe how each tool was placed in the appropriate cell. In the right hand column, convert the sentences into a mathematical shorthand using symbols and mathematical terms.

WRITTEN DESCRIPTION   **SHORTHAND**

CELL A:

CELL B:

CELL C:

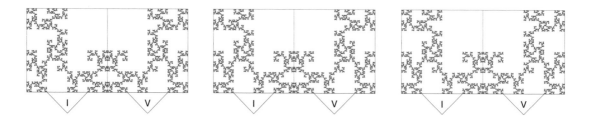

Appendix 12

BLANK TOOL FOR FIND THE CODE

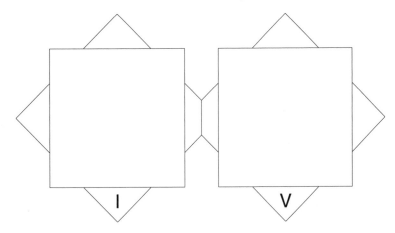

Appendix 13

FIND THE CODE

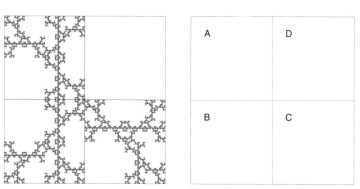

Using the tool at the bottom of the sheet, determine the sequence of transformations, or code, for each cell and record your solutions in the chart below. Discover multiple solutions for each cell. Find solutions that require three moves, then two moves, then one move. It may be possible to find several solutions for each cell.

	3-MOVES	2-MOVES	1-MOVES
CELL A			
CELL B			
CELL C			

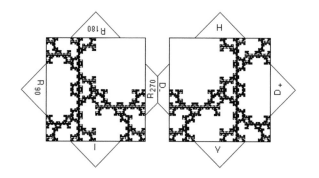

Appendix 14

**FRACTAL WALLHANGING
LIMIT IMAGES**

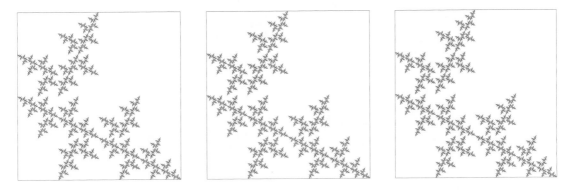

Appendix 15

FRACTAL WALLHANGING GRID SHEET

Appendix 16

PROGRESSION OF A FRACTAL IMAGE

STARTING IMAGE

ITERATION 1

ITERATION 2

ITERATION 3

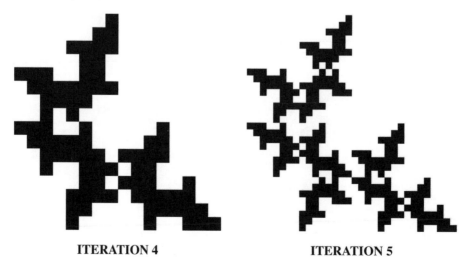

ITERATION 4   **ITERATION 5**

Appendix 17

**FRACTAL WALLHANGING
ITERATION 5 IMAGES**

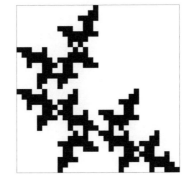 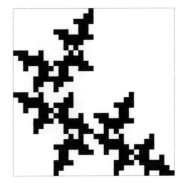

Chapter 10

Other Chaos Games

Sandra Fillebrown

1 Introduction

The Chaos Game has probably been played by countless elementary and high school students over the last five to ten years. Numerous articles and books (Barnsley (1988), Barton (1990), Peitgen, Jurgens & Saupe (1992)) describe the game and provide computer programs that play the game automatically. The rules are simple: Choose three points, call them A, B, and C. Choose a starting point, call it X_0. Generate a sequence of points X_1, X_2, ... by selecting one of the points A, B, or C at random and moving half the distance to that point from your current location. The resulting collection of points will be a picture of the now familiar fractal, the Sierpinski triangle.

One of the appealing aspects of fractals is that their symmetries and complexities can be appreciated by anyone, regardless of the amount of mathematical knowledge they have. The Sierpinski triangle is an especially nice example because its construction can be described by the simple rules of the Chaos Game. The question that arises naturally in this context then, is, can we create other fractals using simple rules like those of the Chaos Game? There are many examples of fractals that have a fairly simple underlying structure. Are there Chaos Game rules that will generate them? Or viewed from the other side, what happens if we change the rules of the Chaos Game? Do we still get fractals? For example, suppose we only move one third the distance instead of one half. Or suppose we move different amounts depending on which point A, B, or C is chosen. Or suppose we have five points instead of three.

These questions were the beginning of a very successful project undertaken by myself and some undergraduates that culminated in bringing a group of 3rd and 4th grade students to our campus to learn about fractals and the Chaos Game during the fall semester of 1994. Every other year I teach an undergraduate course in dynamical systems and fractals in which the students are required to do a project. The course is open to anyone with two semesters of calculus and is generally taken by students majoring in mathematics, computer science, physics, chemistry, or biology. Almost always several students interested in education take the course and I encourage them to consider projects on teaching about chaos or fractals to high school or elementary students. Over the past few years I have been invited to several local schools to talk about fractals and have developed some teaching tools that seem to work well, but I'm always looking for new ideas.

Recently, three students (two math majors, one of whom is getting certified in secondary education, and a computer science major) decided to do their project on Chaos Game rules and how these rules can be used to teach children about fractals. Their project had several parts. First they worked out the mathematics that gives the connection between Chaos Game rules and fractals. Next, the computer science major made some modifications to an interactive computer program I wrote that plays Chaos Games, and the other two students worked with me on teaching strategies. Finally, we invited some elementary school students to come use our computer classroom and learn about fractals. Based on the number of "cools" and "awesomes" we heard while they were here, it was a great success.

This rest of this chapter is broken into three parts: first an explanation of Chaos Game rules, second some examples, and third an explanation of how we taught the 3rd and 4th graders about fractals.

2 Chaos Game rules

Chaos Game rules are associated with a particular type of fractal, the attractor of an iterated function system or IFS, described in Barnsley (1988). Defining an IFS, however, requires some mathematical machinery, while the Chaos Game rules can be described using only very basic geometric constructions. If the symmetries of a fractal can be described as a sequence of scalings, rotations, and reflections, then there is a set of rules for a Chaos Game that will generate the fractal. Conversely, if Chaos Game rules generate a fractal, then that fractal will be the attractor of an IFS.

Briefly, in two dimensions, the attractor of an IFS is the fixed point of a map W which acts on compact subsets of the plane. The map W is defined via a collection of maps $w_1, w_2, \ldots, w_n$, where each w_i is an affine transformation of the plane. Thus, each w_i is defined as mapping points X in the plane via

$$w_i(X) = A_i X + T_i$$

where A_i is a 2×2 matrix and T_i is a translation vector. If C is a compact set, then $w_i(C)$ is what you would expect and

$$W(C) = \bigcup_{i=1}^{n} w_i(C).$$

If each map w_i is a contraction, then so is W. Thus, W will have a unique fixed point, in this case a compact set in the plane. This fixed point is the attractor of the IFS and for certain collections of maps w_i, this set will be a fractal. See Barnsley (1988) for the details of the appropriate metrics, etc.

To see the attractor one can employ the so called Random Iteration Algorithm. Choose an initial point X_0. Then generate a sequence of points $X_1, X_2 \ldots$ by selecting one of the maps w_i at random and applying it to the current point. The resulting set of points will produce a picture, the attractor of the IFS. Details as to the manner in which the collection of points $\{X_k\}$ approximates the attractor can be found in Barnsley (1988). Some examples of such fractals along with their maps w_i are given in Figures 1–4. Many of the fractals found in the popular literature are generated in just this way.

Let us now restrict ourselves to maps w_i that are composed only of scalings, rotations, reflections and translations. In this case the matrices A_i associated with each map can be written as the product of three types of matrices. The map must contain at least one scaling of the form

$$\begin{pmatrix} s & 0 \\ 0 & s \end{pmatrix},$$

where $0 < s < 1$. (A generalization would allow different scaling factors in each coordinate, but this leads to Chaos Game rules with less visually pleasing attractors.) It may contain a reflection of the form

$$\begin{pmatrix} -1 & 0 \\ 0 & 1 \end{pmatrix},$$

or

$$\begin{pmatrix} 1 & 0 \\ 0 & -1 \end{pmatrix},$$

and it may contain a rotation of the form

$$\begin{pmatrix} \cos\theta & -\sin\theta \\ \sin\theta & \cos\theta \end{pmatrix}.$$

Note that each of these transformations is defined with respect to the origin as is any translation T_i associated with w_i.

Now suppose a particular fractal is associated with a collection of maps that consist only of scalings, reflections, rotations, and translations. The Chaos Game rules for such a fractal can be discovered by recasting the definitions of the maps w_i so that they do not explicitly depend on an underlying coordinate system, as do any scalings, reflections, rotations, and translations that are done with respect to the origin. Since each map of an IFS is a contraction, it will have a unique fixed point (a single point in the plane). Maps that are just scalings, rotations, and translations can all be described easily in terms of this fixed point with no reference to a coordinate system and maps that include a reflection can be described in terms of the fixed point along with some notion of orientation, i.e., what is vertical or horizontal. Thus, the Chaos Game rules come from a simple change of coordinates for each map from the origin to its fixed point.

To see how this works, suppose w is any affine transformation that is a contraction. Let P be its fixed point and let w map points X via

$$w(X) = AX + T.$$

Then $T = P - AP$ and w can be rewritten in terms of P as

$$w(X) = AX + (P - AP)$$
$$= P + A(X - P).$$

Thus, $w(X)$ can be described as the result of modifying the vector from P to X by the matrix A and adding it to the vector P. In this formulation, any translation component of w is absorbed and all other transformations can be described with respect to the fixed point P instead of the origin. Let us now interpret each of these transformations (scaling, rotation and reflection) geometrically to produce rules for a new Chaos Game.

Suppose A is a scaling by a factor of s so that A is given by

$$\begin{pmatrix} s & 0 \\ 0 & s \end{pmatrix}.$$

Then the rule is just move $1 - s$ of the distance from where you are toward the fixed point P. (Since w is a contraction, we know that $0 \le s < 1$.) Thus, if you are at the point X, your new point will be $X + (1 - s)(P - X)$. This is true because

$$w(X) = P + A(X - P)$$
$$= P + s(X - P)$$
$$= X + (1 - s)(P - X).$$

Next, suppose A is a rotation of θ degrees counterclockwise. Then A has the form

$$\begin{pmatrix} \cos\theta & -\sin\theta \\ \sin\theta & \cos\theta \end{pmatrix},$$

and the rule for this point would just be to rotate around P by θ. To see why this is so, note that rotating a point X around P by θ is done with the transformation

$$P + \begin{pmatrix} \cos\theta & -\sin\theta \\ \sin\theta & \cos\theta \end{pmatrix}(X - P),$$

which is exactly $w(X)$. (Geometrically, you are just shifting the coordinate system by P, rotating about the origin, and then shifting the coordinate system back by P.) Finally, if A is a reflection, then the rule will involve reflecting the point X across a vertical or horizontal line through P. (Note that this transformation requires an underlying fixed orientation, while the others do not.) Thus, if the matrix A were just

$$\begin{pmatrix} -1 & 0 \\ 0 & 1 \end{pmatrix},$$

i.e., a reflection across the vertical, then the rule would be to reflect X across a vertical line through P. In general, a reflection across a vertical line through a point P is given by

$$P + \begin{pmatrix} -1 & 0 \\ 0 & 1 \end{pmatrix}(X - P),$$

which, again, is exactly $w(X)$. (Reflections across the horizontal are completely analogous.)

To summarize, suppose we have an IFS with N maps w_i and suppose each map has an associated fixed point P_i. Suppose each map w_i can be written in the form

$$w_i(X) = P_i + A_i(P_i - X),$$

where A_i is the product of a scaling by s in both coordinates and then possibly a horizontal or vertical reflection or a rotation by θ. Then we can associate with each fixed point P_i a Chaos Game rule, consisting of moving $1 - s$ of the distance toward P_i, then possibly reflecting across a vertical or horizontal line through P_i, and then possibly rotating around P_i by θ. (Since reflections and rotations do not commute, the order is important.) To play the Chaos Game, select an initial point X_0, and generate X_{i+1} from X_i by choosing one of the N fixed points at random and applying the associated rule to X_i.

Note that we can consider this from the other direction as well. We can choose N points P_i and associate with each point a rule that consists of moving a certain fraction of the distance towards the point and then possibly reflecting or rotating. These rules will define a Chaos Game and also an IFS.

3 Some examples

In the case of the Sierpinski triangle, there are three maps, each of which is a scaling by $1/2$ followed by a translation. The particular translations chosen determine the shape of the triangle: the three fixed points are the vertices of the triangle. See Figure 1. The three transformations reduce to

$$w_i(X) = P_i + \tfrac{1}{2}(X - P_i),$$

where P_i is just the fixed point of the map w_i, one of the vertices of the triangle. Thus, we get the original rules of the Chaos Game.

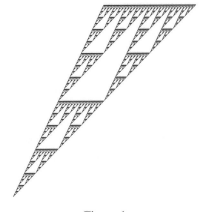

Figure 1

As another example, consider the Chaos Game with three fixed points P_1, P_2, and P_3 and the following rules:

P_1: Move $\frac{1}{2}$ the distance towards P_1.

P_2: Move $\frac{1}{2}$ the distance towards P_2, then reflect across the vertical.

P_3: Move $\frac{1}{2}$ the distance towards P_3, then rotate $90°$.

Depending on where the fixed points are placed, the Chaos Game will generate fractals that are visually quite different. Figures 2 and 3 are two such examples. Can you locate the fixed points?

Figure 2

As a final example, the following Chaos Game rules generate the dragon fractal (Figure 4). Choose two fixed points P_1 and P_2 with the rules:

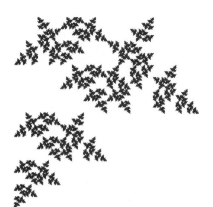

Figure 3

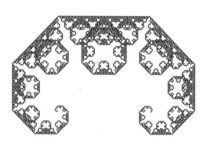

Figure 4

P_1: Move $1 - \frac{\sqrt{2}}{2} \approx \frac{3}{10}$ the distance towards P_1, then rotate by $45°$.

P_2: Move $1 - \frac{\sqrt{2}}{2} \approx \frac{3}{10}$ the distance towards P_2, then rotate by $-45°$.

As with the Sierpinski triangle, the location of the fixed points does not significantly alter the appearance of this fractal. For Figure 1 the $w_i \binom{x}{y} =$

$$\begin{pmatrix} 1/2 & 0 \\ 0 & 1/2 \end{pmatrix}\begin{pmatrix} x \\ y \end{pmatrix}, \quad \begin{pmatrix} 1/2 & 0 \\ 0 & 1/2 \end{pmatrix}\begin{pmatrix} x \\ y \end{pmatrix} + \begin{pmatrix} 1/4 \\ 1/2 \end{pmatrix},$$
$$\text{and } \begin{pmatrix} 1/2 & 0 \\ 0 & 1/2 \end{pmatrix}\begin{pmatrix} x \\ y \end{pmatrix} + \begin{pmatrix} 1/2 \\ 1/2 \end{pmatrix}.$$

For Figure 2 the $w_i \binom{x}{y} =$

$$\begin{pmatrix} 1/2 & 0 \\ 0 & 1/2 \end{pmatrix}\begin{pmatrix} x \\ y \end{pmatrix}, \quad \begin{pmatrix} -1/2 & 0 \\ 0 & 1/2 \end{pmatrix}\begin{pmatrix} x \\ y \end{pmatrix} + \begin{pmatrix} 1 \\ 0 \end{pmatrix},$$
$$\text{and } \begin{pmatrix} 0 & -1/2 \\ 1/2 & 0 \end{pmatrix}\begin{pmatrix} x \\ y \end{pmatrix} + \begin{pmatrix} 1/2 \\ 1/2 \end{pmatrix}.$$

For Figure 3 the $w_i \binom{x}{y} =$

$$\begin{pmatrix} 1/2 & 0 \\ 0 & 1/2 \end{pmatrix}\begin{pmatrix} x \\ y \end{pmatrix}, \quad \begin{pmatrix} -1/2 & 0 \\ 0 & 1/2 \end{pmatrix}\begin{pmatrix} x \\ y \end{pmatrix} + \begin{pmatrix} 1/2 \\ 1/2 \end{pmatrix},$$
$$\text{and } \begin{pmatrix} 0 & -1/2 \\ 1/2 & 0 \end{pmatrix}\begin{pmatrix} x \\ y \end{pmatrix} + \begin{pmatrix} 1 \\ 1/2 \end{pmatrix}.$$

For Figure 4 the $w_i \binom{x}{y} =$

$$\begin{pmatrix} 1/2 & -1/2 \\ 1/2 & 1/2 \end{pmatrix}\begin{pmatrix} x \\ y \end{pmatrix} \quad \text{and} \quad \begin{pmatrix} 1/2 & 1/2 \\ -1/2 & 1/2 \end{pmatrix}\begin{pmatrix} x \\ y \end{pmatrix} + \begin{pmatrix} 1 \\ 1 \end{pmatrix}.$$

4 Teaching elementary school children

To teach elementary school students about Chaos Games, we used a combination of low-tech and high-tech strategies. We used paper and pencil, rulers and protractors, and transparencies to explain the possible transformations. Next we let the students experiment with the computer program, creating arbitrary Chaos Game rules and looking at the results. Then we spent some more time with the transparencies explaining about fixed points and challenging them to reproduce a specific fractal on the computers. And finally, we let each of them design their own fractal, of which we gave them a printed copy to take home.

The computer program is a mouse driven DOS program that allows the user to change the Chaos Game rules and then produces a picture of the associated fractal. You can add or delete points or move an existing point. For each point, you can modify the rules by selecting a new scaling factor or a new rotation factor and by toggling between including or not including a horizontal or vertical reflection. The program includes several other options such as recentering, zooming in or out, using different colors for all the points obtained from each fixed point, and saving the information for a particular fractal. All of the children found the program easy and fun to use.

To teach about the different types of transformations, we used a more traditional approach. Each student had a copy of ten different fractals generated by Chaos Game rules. The students were split into four groups and were given rulers, protractors, and a packet of transparencies. Each packet of transparencies contained at least one picture that was a reduced-size version of each of the fractals. The reductions were chosen based on the scaling factors corresponding to each fractal. The students first learned how to identify the smaller copies of the whole fractal. The easy ones (the ones without too many rotations or reflections or the ones that weren't overlapping) they could do by inspection; for the more difficult ones they used the transparencies. By sliding, turning and flipping over the transparencies, they were able to isolate the different pieces of the fractal and also determine the sequence of rotations and reflections that generated each one. They used the rulers to measure different lengths on the fractals to find the scaling factors. These activities are not too difficult, and (aside from each wanting their own set of transparencies!) the children enjoy this activity. Each group was helped by me or one of the undergraduates.

The most difficult concept that we tried to teach was that of fixed points. Again, we used the transparencies to help the

Figure 5: Kids learning to draw fractals.

students visualize what was going on. They started by identifying and marking the point on the fractal that they thought was the fixed point. Then they marked the corresponding point on the transparency of the reduced copy. When they moved the transparency (rotating or reflecting as necessary) over the piece of the fractal they were working on, they could see if the points ended on top of each other, i.e., the point stayed fixed. All of the students discovered fairly quickly where the fixed points would be for pieces that involved only scalings. Many were able to develop a strategy for finding the fixed points of pieces that had only scaling and rotations. When asked to explain their ideas, they usually described an approach of trying to find the end of the spiral. Pieces that had reflections were the most difficult and the few students who were able to find the fixed points for these used a *hotter–colder* technique, i.e., if the two points (the one on the fractal and the one on the transparency) were closer together than

their last guess, they were on the right track and if they were farther apart, they were going in the wrong direction. Regardless of the amount of success they had, the students seemed fascinated, not frustrated, and if we had not run out of time, perhaps many more would have caught on.

5 Conclusion

The simplicity of the Chaos Game rules make them an ideal starting place for teaching about fractals. Students as young as third and fourth graders can understand the rules and some of the properties of the fractals related to them. At the other end of the spectrum, explaining the connections between Chaos Game rules, the fractals they produce and the IFS codes makes a good project for undergraduates. Putting the two groups together provided a wonderful experience for all.

Chapter 11

Creating and Teaching Undergraduate Courses and Seminars in Fractal Geometry: A Personal Experience

Michel Lapidus[1]

Since moving to the University of California at Riverside in 1991, I have created and taught many undergraduate and graduate courses or seminars in Dynamical Systems, Fractal Geometry, and Chaos. Far from distracting from my own research program, these pedagogical and the ensuing mentoring activities have proved to be very stimulating and rewarding.

I will stress here the aspects most relevant to the major theme of this book and discuss mostly the undergraduate courses and seminars. However, when appropriate, I will briefly point out the connections between the undergraduate and graduate aspects of this fulfilling experience.

Here is a brief plan of this chapter. After presenting a brief overview of the project, I provide in Appendix 1 a representative syllabus for one of my courses (Honors 23E), along with a short bibliography given to the students at the beginning of the class. Then, in Appendix 2, I give more detailed information about the undergraduate courses or seminars themselves, as well as about some of the students who have taken them. (Appendix 2 is a slightly updated version of the final report of a Teaching Excellence Grant which I was awarded a few years ago by the University of California in order to develop these courses and seminars at UC Riverside.) Finally, a more extensive (but certainly not exhaustive) bibliography is provided at the end of the paper. Also included at the end of the bibliography is a selection of instructional films which I have used in my courses.

I hope this brief narrative of my own experience will be helpful to some of the readers and will inspire them to create or further develop their own courses on the fascinating subject of fractal geometry and chaos.

It is a pleasure to thank here Professor Benoit B. Mandelbrot for encouraging me to write this paper, and for our initial discussions about the first course on fractal geometry, *Fractal Geometry with Application* (MATH 141A & B) that I was about to create at the University of California, Riverside, during the 1990–91 academic year when I was a visiting professor at Yale University upon his kind invitation.

1 Dynamical systems and fractal geometry

In September 1991, the Office of Instructional Development of the University of California awarded me a Teaching Excellence/Major Instructional Development Grant to develop new courses or seminars on Fractal Geometry and Dynamical Systems during the 1991–92 and 1992–93 academic years. At the undergraduate level, in addition to the course *Fractal Geometry with Applications* (MATH 141A & B), I created and taught a *Junior Seminar in Dynamical Systems* (MATH 191F, offered for the first time in the Spring of 1993), and a *Sophomore Honors Seminar on Fractal Geometry and Chaos* (Honors MATH 22M, offered for the first time in the Spring of 1994).

At the graduate level, I created and taught a *Seminar on Fractal Geometry and Dynamical Systems* (MATH 260-7, with partial emphasis on complex dynamics) in the Spring

[1] Work supported by a Major Instructional Improvement Project/Teaching Excellence Grant from the Office of Instructional Development of the University of California, as well as by research Grants from the National Science Foundation under contracts DMS-8904389 (transferred to the University of California as DMS-9196085), DMS-9207098 and DMS-9623002.

of 1991. (This is how I was first put in contact with two of my recent Ph.D. students, Ms. Cheryl-Ann Griffith and Ms. Christina He.) This was followed in 1992–93 by a year-long *Graduate Seminar in Spectral and Fractal Geometry*. During every academic year since then, I have offered a year-long *Graduate Seminar in Mathematical Physics and Dynamical Systems*. This seminar is often attended by upper-division undergraduate students who first became interested in the subject by attending the aforementioned undergraduate courses or seminars. Frequently, these students have gone on to graduate school in Mathematics and have been rather successful in their studies.

2   Honors Program

Since the academic year 1993–94, I have been very actively involved in the Honors Program of the University of California, Riverside. I have created a Sophomore Honors Seminar on Fractal Geometry and Chaos (Honors MATH 22M) which I offered for the first time in the Spring of 1994 and which I taught until the 1997–98 academic year. According to the then director of the Honors Program, Professor Alex Rosenberg, this course was very successful. It had for prerequisite only one variable calculus (MATH 9A & 9B) and was attended by thirteen students (including two Freshmen) majoring in Mathematics, Physics, Biology, Pre-Medicine, Information Sciences, Drama, and Philosophy. In my course, I strongly encourage students' participation and creativity, in particular in the form of projects prepared and presented during the seminar. This subject lends itself very well to this endeavor. Probably one of the greatest challenges in preparing for the lectures is to take into account the diverse background and levels of my students. Partly in response to this problem, I attempt to cater to their individual needs by many one-on-one conversations. (A sample of my syllabus is in Appendix 1.)

Each of my students presented a course project. Some were particularly noteworthy, like that of Erin Pearse, an exceptional student (now my advisee and future Ph.D. student) on *Phase Space and the Ergodic Hypothesis*, that of Prashant Phatak (a Biology major) on the *Chaotic Solutions of van der Pol Equations and Cardiac Arrhythmia*, and that of Wesley Stevens on the *Design of Plants and Fractals by means of L-Systems*. (A broader sample of the titles of my students' projects is provided in Appendix 2.)

My students were very enthusiastic about the course and I believe that they have learned a lot of mathematics (and physics) in the process of understanding the material and preparing their projects.

During the academic year 1993–94, I was also a Faculty Mentor for the Honors Colloquium. In that capacity, I had to help a group of fifteen Freshman Honors students prepare throughout the year for the presentation (in May 1994) of a two-hour long Colloquium on Chaos and Complexity in front of all the other (100 or so) Freshmen from the Honors Program. Mr. Pearse, as well as several other students from the Sophomore Honors Seminar, participated in this colloquium. Although it was often quite involved, their presentation—which consisted of physical experiments, mathematical explanations of various phenomena, and of examples of applications to economics and biology—was generally well received by the audience.

In as much as possible, in these courses or seminars I also try to provide my students with a suitable historical perspective placing the subjects of fractal geometry and chaos within the broader context of mathematics. Besides the purely internal, philosophical or aesthetic motivations, I stress the motivations coming from the observation of nature. For example, clouds, mountains, trees, lungs, blood vessels, and coastlines are fractals. During the course, I discuss at some length several applications of the subject to physics, chemistry, engineering, computer graphics, and biology.

During the first two-thirds of the seminar/course, I provide the foundations for some of the basic aspects of the subject, and then my students and I alternate lectures, as they present their research.

I wish to stress that the syllabus included in Appendix 1 is only indicative of the material covered in a typical course or seminar. Different topics are emphasized, depending on the students' interests, their mathematical sophistication and their majors. Of course, in the two-quarter sequence, aimed at more advanced upper-division students, I go beyond the enclosed syllabus, and the topics are approached in a different manner than in the lower-division courses or than even in the Honors seminars.

I often make use of the Socratic method. Further, especially in the lower level courses I use suitable examples and analogies to provide enough intuition so my students eventually will be able to guide themselves through this somewhat disorienting but always attractive and surprising subject.

Finally, let me mention that towards the end of the course, I often briefly discuss some aspects of my own research on fractal and spectral geometry and on the vibrations of fractal drums. See, for example, Lapidus (1991), Lapidus (1992), Lapidus (1993), Lapidus (1994), Lapidus (1995), Lapidus (1997), Lapidus & Pomerance (1993), Lapidus & Pomerance (1996), Lapidus & Maier (1995), Kigami & Lapidus (1993), He & Lapidus (1996), He & Lapidus (1997), Lapidus & Pang (1995), Lapidus, Neuberger, Renka, & Griffith (1996), Griffith & Lapidus (1997), and Lapidus & van Frankenhuysen (1999a,b). Since it would be futile to attempt to communicate the more mathematically sophisticated side of my research to my undergraduate students, I emphasize some of the physical motivations. These consist in trying to understand how fractal shapes arise in nature. In this, I also present some of the computer graphics work that my collaborators and I have produced in order to tackle aspects of this question. See especially Lapidus, Neuberger, Renka, & Griffith

(1996) and Griffith & Lapidus (1997), along with the closely related theoretical work Lapidus & Pomerance (1993). Also see the earlier physical and numerical work by Sapoval and his collaborators, as discussed in Sapoval (1989) and the relevant references therein or in Lapidus, Neuberger, Renka, & Griffith (1996).

The students are always very interested by the computer graphics of fractal drums (from Lapidus, Neuberger, Renka, & Griffith (1996)) and want to know more about the way they were produced, their interpretation, and relevance. In my last Honors Seminar, I gave a very brief and elementary introduction to the notion of complex dimension that I have developed in recent joint work (at first heuristically in Lapidus & Pomerance (1993), Lapidus & Maier (1995), Lapidus (1992), Lapidus (1993), and then rigorously in Lapidus & van Frankenhuysen (1999a) and especially in Lapidus & van Frankenhuysen (1999b)).

This is only intended to show that mathematics, in its many disguises, is alive and constantly progressing. The subjects of fractal geometry and chaos are particularly concrete and potent examples of this undeniable, but often ill-recognized, fact.

Appendix 1

COURSE OUTLINE FOR HONORS 23E
Honors Sophomore Seminar HNPG 23E
MATHEMATICS:
FRACTAL GEOMETRY AND CHAOS

Instructor: Dr. Michel L. Lapidus

Week 1: What is a fractal?
Self-similarity in nature.

Week 2: A basic example of self-similar fractal:
The ternary Cantor set.

Week 3: Fractal coastlines and mathematical fractal curves
Physical vs. mathematical fractals.
(Length scale and cut-off.)

Week 4: Introduction to the notion of fractal dimension and measure.

Week 5: Further examples of self-similar fractals:
The Koch snowflake curve; the Sierpinski gasket and carpet; Peano's curve.

Week 6: Fractals in biology (e.g., lungs, blood vessels, trees), geology (clouds, mountains, rivers) and physics (e.g., percolation).

Week 7: The Chaos Game and the Sierpinski gasket.

Week 8: Chaotic dynamical systems and the end of determinism.

Week 9: Period doubling route to chaos.

Week 10: Complex dynamical systems: an introduction to Julia sets and the Mandelbrot set.

READING LIST FOR HONORS 23E

Required Reading:

Mandelbrot (1982), Peitgen, Jurgens, & Saupe (1992a) (Especially Chapters 1–4, 10–11, 13–14).

Recommended Reading List:

Peitgen & Richter (1986), Schroeder (1991), Barnsley (1988), Devaney (1990).

Course Requirements:

Students need to present a talk in the seminar and to write a term paper based on a small research project connected with the seminar. Regular attendance and active participation are expected.

Appendix 2

FINAL REPORT

Major Instructional Improvement Project/Teaching Excellence Development of New Undergraduate Courses and Seminars on Fractal Geometry and Dynamical Systems, with Application, 1991–1993.

University of California, Riverside

Michel L. Lapidus
Professor of Mathematics
October 1993

1. **Report on MATH 141A & B (New Course on "Fractal Geometry with Applications")**

As was my intention all along, I have aimed the course *primarily at undergraduate students*, even though several graduate students were also attending. It was attended by ten students and one (Mathematics) Faculty member, Professor Vernon Howe, of Loma Linda University. (All my students from Fractal Geometry with Applications A took Fractal Geometry with Applications B, except that during the second quarter, one student has audited the course instead of taking it for credit.) My students had majors ranging from Mathematics, Physics, Computer Science and Engineering to Plant Physiology and Philosophy. Hence my course was truly *cross-disciplinary*. I have assumed very little background at the beginning (besides a calculus sequence and sufficient mathematical maturity) but was able to cover some more advanced material at the end of the first quarter. (The basic examples of self-similar fractal that I have covered in depth in Fractal Geometry with Applications A have been very helpful in order to understand the more delicate and abstract concepts that were covered in Fractal Geometry with Applications B.)

During the course, I have presented several films, including one featuring extensive interviews of Professors Edward Lorenz and Benoit Mandelbrot about the nature of chaos and of fractal geometry (Peitgen, Jurgens, Saupe & Zahlten (v1990)). Depending on the interests of the students and the topics covered in the course or the seminar, these instructional films are selected, in particular, from the list **Instructional Films** given at the end of the references.

Each of my students has completed a project during the course. Some of their projects were quite impressive, like that of Zachary Mason (a Junior and an exceptional student whom I have nominated for a UC Presidential Undergraduate Fellowship at UCR) on the *Four-Dimensional Mandelbrot Set* and that of Sybille Thamm on the *Period Doubling Route to Chaos*, as well as those of David Dixon on *Ising Models on Fractal Lattices* and of Wei-Yun Yu on *Newton's Method for Complex Polynomials*. In MATH 141B, they have given an oral presentation of their project and also worked on a related project.

My students were very enthusiastic about the course and I believe that they have learned a lot of mathematics (and physics) in the process of understanding the material and of working out the (numerous) homework problems.

Finally, let me mention that when I asked my students who would be interested in attending the (student-oriented) courses on Fractal Geometry and Complex Dynamical Systems organized by Professor Devaney during the summer at Boston University under the sponsorship of the National Science Foundation, almost all of them were eager to participate in this PROMYS (Program in Mathematics for Young Scientists) program. This is the best reward that a teacher can receive from his students.

Here is some information about several of my students since they have taken MATH 141A & B:

   Zachary Mason was awarded the UC Presidential Undergraduate Fellowship but, unfortunately, could not accept it because he was moving to Harvey Mudd College where he has pursued work related to his project in Fractal Geometry with Applications. He is presently applying for graduate school at MIT and Columbia, among other schools. (Of course, I wrote a very strong letter supporting his application.)

   Sybille Thamm is now back in Germany and is taking her teaching credentials. Before returning to Germany, she told me that she was planning (as part of her training for high school or college teaching) to develop a curriculum for a course on Fractal Geometry inspired

by my course Fractal Geometry with Applications A & B.

   David Dixon is a Ph.D. student in Physics at UCR working on Dynamical Systems. (I am a member of his Ph.D. committee and he went on to take all the graduate seminars on fractal geometry and dynamical systems that I have offered since then on the subject.)

   Robert James Miller, another very bright undergraduate student and now a Mathematics major, has also attended my graduate seminar on Dynamical Systems in Spring, 1992, and is presently taking my new undergraduate seminar *Fractal Geometry and Dynamical Systems* (MATH 191F). He plans to go on to graduate school in Mathematics and do research in fractal geometry and dynamical systems. (Actually, he told me that he had changed his major from Computer Science to Mathematics because of his enthusiasm for the material that he had learned in Fractal Geometry with Applications.)

   Professor Vernon Howe has been giving a course on fractal geometry at Loma Linda University, and many expository lectures in local high schools and colleges on material related to Fractal Geometry with Applications A. He has told me that attending my courses, Fractal Geometry with Applications A & B, has been very helpful in this endeavor.

2. Besides the new undergraduate seminar Fractal Geometry and Dynamical Systems (which will be discussed below), I have also created graduate seminars in Dynamical Systems or Fractal Geometry (MATH 260-4, Spring 1992) and (MATH 260-7, Fall 1992, Winter & Spring 1993, to be continued during the academic year 1993–94). MATH 260-4 was an introductory seminar which, although primarily aimed at graduate students, was also attended by undergraduate students, including Robert James Miller. My two current Ph.D. students, Cheryl Griffith and Christina He, were first attracted to the field by taking MATH 260-4. Christina He, an extremely bright student, has learned the subject so quickly that she is presently my Teaching Assistant for my undergraduate seminar, MATH 191F. Cheryl Griffith will probably assist me in computer graphics programming during future courses. She is also a very capable student and should become an outstanding experimental mathematician.

It is probably fair to say that, although fractal geometry had never been taught at UCR before I created these graduate and undergraduate courses and seminars, it has now had a very significant impact in the students' mathematical life. These courses have helped build bridges between seemingly unrelated topics or disciplines, as well as warm working relationships between graduate and undergraduate students, united by a common love for this subject.

Remark. (a) Since his freshman year, Mr. Robert James Miller has taken from the author a number of courses and Honors seminars related to Fractal Geometry and Dynamical Systems. He graduated from UCR about two years ago with a major in Mathematics and a minor in Philosophy. Upon being conferred his undergraduate degree, he received a well-deserved campus-wide Creative Research Award for his research projects in my courses and seminars. Being a very original and well-rounded student, he has taken some time off to dedicate himself to other creative and artistic endeavors (in particular, dance and writing). As far as I know, he now lives in a community in Northern California and still plans to go on to graduate school in Mathematics at some later point. He continues reading a lot about fractals and dynamical systems and comes to see me from time to time to seek my advice on these and related matters.

(b) Mr. Erin Pearse is another very creative and unusually bright student whom I have known since his freshman year at UC Riverside. Together with fourteen other freshmen, he has also participated in an Honors Freshman Colloquium on Complex Systems and Fractal Geometry, for which I was the Faculty Mentor. (At the end of the academic year, this group of students had to present an hour and a half colloquium to all the other Honors students.) In addition, he has taken my Sophomore Honors Seminar in *Fractal Geometry and Chaos* (Honors 23E) and related courses from me. He is also a Mathematics major with a minor in Philosophy (apparently, a very successful academic combination). He has been my advisee for several years. Over the last two years, he has worked under my supervision on an Honors Undergraduate Thesis entitled *Universality of the Sierpinski Carpet*, which he defended in June 1998. (See Pearse (1998), which discusses in detail the original proof (Sierpiński (1916)) of the fact that every Jordan curve can be embedded homeomorphically in the Sierpinski carpet, as well as its extension by Karl Menger (reprinted on pp. 103–117 of Edgar (1993)) to higher-dimensional analogues of the carpet, such as the Menger sponge.) For this and earlier work under the author's guidance, he was awarded a campus-wide Creative Research Award in the Spring of 1998. He has also received the Phi-Beta-Kappa Award from the Honors Program upon graduation in June 1998. For financial reasons, he is now taking a one year break before beginning his graduate studies in Mathematics during the next academic year. He clearly is an excellent prospect as a Ph.D. student. For his dissertation, he wants to continue his research on Fractal Geometry and Dynamical Systems.

Erin Pearse was awarded the Chancellor's Distinguished Fellowship Award for which I had nominated him, and that he would definitely return to the University of California, Riverside, in the Fall of 1999 in order to pursue his graduate studies in Mathematics and to work on his Ph.D. under my supervision. This fellowship will support him throughout his graduate studies and enable him to spend considerably more time on his research by relieving him, in particular, of his teaching duties for two full years.

Remark 2. Cheryl Griffith went on to join the author's research project (joint with John Neuberger and Robert Renka) on Computer Graphics and the Vibrations of the Koch Snowflake Drum, which gave rise to the publications Lapidus, Neuberger, Renka, & Griffith (1996) and Griffith & Lapidus (1997), building, in particular, on the earlier physical (Sapoval (1989)) and mathematical (Lapidus & Pang (1995)) works. Christina He has worked with the author on more theoretical questions related to Generalized Minkowski-Bouligand Content or Dimension and the Vibrations of Fractal Drums and Strings, which gave rise to the long memoir He & Lapidus (1997) (announced in He & Lapidus (1996)) and building, in particular, on the earlier work of the author and his collaborators (Lapidus (1991), Lapidus (1992), Lapidus (1993), Lapidus (1995), Lapidus & Pomerance (1993), Lapidus & Pomerance (1996), Lapidus & Maier (1995)), now further extended in Lapidus & van Frankenhuysen (1999a,b). Both Ms. Griffith and Ms. He have received their Ph.D. in Mathematics in June 1996 under the author's supervision.

3. **Preliminary Report on MATH 191F (New Undergraduate Seminar on Fractal Geometry and Dynamical Systems)**

I have created MATH 191F, entitled Undergraduate Seminar on Fractal Geometry and Dynamical Systems in the Fall of 1992. (It was accepted by the committee on courses in the Winter of 1993 and I have offered this seminar for the first time during the Spring of 1993. (Please see the enclosed course announcement.) The aim of the seminar is to initiate undergraduate students to mathematical and pluridisciplinary research, through a gradual exposure to some of the main concepts of the theory of dynamical systems and fractal geometry.

The seminar was attended by twelve students with various backgrounds (mathematics, physics, industry, biology). Each student has chosen a topic for an oral presentation and a written project with an emphasis on mathematics, physics, mathematical biology, or computer graphics. Here is a sample of their projects:

Henrik Thistrup: *Complex Dynamics, Julia Sets and the Mandelbrot Set.*

Robert James Miller: *An Introduction to Fractal Dimensions.*

Catherine Huang: *Self-Similar Fractals.*

Laurel Langford: *Topological Universality of the Sierpinski Carpet.*

Terry Bauer: *An Introduction to Percolation Theory.*

Darren Anderson: *Brownian Motion and Random Fractals.*

Philipp Turner: *Chaotic Dynamical Systems and Feigenbaum's Bifurcation Diagram.*

Karen Mifflin: *The Chaos Game and Symbolic Dynamics.*

Michelle McBride & Viviane Frazee: *The Algorithmic Beauty of Plants; Theory and Computer Graphics Experimentation.*

I have been working for a long time in order to prepare for the Junior Seminar in Dynaical Systems (as well as future courses or seminars on related subjects). In particular, I have used my Teaching Excellence Grant for (i) purchasing (or making, with the help of the Media Center) color slides, (ii) purchasing instructional videos directly related to fractal geometry or dynamical systems, (iii) computer software needed for the courses (Mathematica, software on fractals, graphics software, etc.), (iv) the salary of Diana Nguyen, a programmer/assistant to help with the many tasks related to the preparation of the courses, especially the first time they are offered.

All in all, I would say that this has been an extremely positive experience both for my students and myself. I am grateful for the support provided by the Teaching Excellence Grant and I do think that it made a real difference in my ability to attract students to these fascinating subjects and expose them to a rich array of results and techniques that should be useful in their future careers and should further their intellectual development.

Finally, let me mention that very recently, I was asked to join as a Co-Principal Investigator/Major User, a Proposal for the National Science Foundation Multi-User Biological Equipment and Instrumentation Resources, aimed at acquiring several items (of the order of $200,000) to augment the Academic Computing Graphics and Visual Imaging Laboratory [ACGVIL] (now called the Center for Visual Computing, CVC).

I was told by Dr. Ware that in addition to my present research support by the National Science Foundation (on related subjects), my current Major Instructional Development Grant/Teaching Excellence Grant from UCR and my direct involvement in curriculum development for courses on fractal geometry and dynamical systems directly relevant to the use of computer graphics should be extremely useful in securing the above NSF equipment grant. Moreover, even though the ACGVIL is primarily a research facility, I was told that some of my students (and of the students of other investigators) in the new courses and seminars on fractal geometry and dynamical systems that I created over the last few years, will be allowed to use the lab while working on special computer graphics projects related to the courses[2]. Hence, albeit indirectly, my present Teaching Excellence Grant may help a wide array of selected students (in Biology, Biochemistry, Botany, Psychology, Chemistry, Neuroscience, Physics and Mathematics) gain access to state of the art computer graphics facilities during their studies at the University of California, Riverside.

We close this paper by adding two final comments written since the completion of this report.

Remark 3. The enrollment in Honors undergraduate seminars is usually limited to fifteen students. However, the instructor has the possibility of increasing this limit. My own Honors seminars have had between six and eighteen students. In the latter case, I have felt that I could not devote enough individual attention to my students and that there was no longer enough time for a thorough presentation of their projects during the course. (An Honors seminar is limited to one quarter. I might not have encountered the same problem if the seminar had lasted a semester rather than a quarter.) My own experience suggests that the ideal number of students for an Honors-type seminar lies between ten and twelve students.

Remark 4. For a little over a year, I have embarked upon a new pedagogical adventure, namely, the creation of an Undergraduate Courses and Seminars in Mathematical Biology, for which I was awarded another Teaching Excellence Grant by the University of California (September 1997–June 1999). I have offered my first Honors Seminar on this subject in the Spring of 1998 and I intend to create a more advanced (upperdivision) undergraduate course in Mathematical Biology in the near future. The response from the students has been extremely encouraging. I have used a format similar to the one described above, except, of course, that the material taught was rather different. There is, however, a nontrivial overlap between the two subjects, especially from the point of view of dynamical systems and chaos. (In the future, I also intend to have some of my students explore the fractal geometry of the lungs, the blood vessels, or the heart.) In fact, several of my best undergraduate students eagerly want to take my courses or seminars in both subjects. In a few years' time, I hope to be able to report in a different venue on this new experience and on its relationship with the present one.

---

[2]Added note (1998): Due in part to space and financial limitations, this has only been possible in a more restricted way than originally anticipated; that is, mostly for undergraduate (and especially graduate) students directly involved in a project connected with the author's research program. Several undergraduate students did benefit from it, however, either directly or in a more indirect fashion, during the preparation of their research project or of their Honors thesis under the author's guidance. In any case, all the students had access to suitable computing facilities, albeit less sophisticated than those of the Center for Visual Computing.

Chapter 12

Exploring Fractal Dimensions by Experiment

Ronald Lewis

1 A mathematics course for modern times

People are interested in fractals and chaos. Over the last few years the questions they ask upon finding that I teach these subjects have been of this sort:

1) "What are they anyway?"

2) "Are they good for anything?" (Often with an insinuation that they probably have no practical value.)

3) "How do they relate to the present curriculum?"

Most people think of fractals as Mandelbrot sets. These are the very beautiful, often highly colored, sometimes animated, computer images that have appeared in many books, popular magazines, and on television. And so they are! But fractals are much more than pretty pictures.

In 1987, a small group of my students began participating in a pilot project at Laurentian University in Sudbury, Ontario, Canada. The project was conducted by Dr. Brian H. Kaye, who was just completing his book, Kaye (1989). The students were initially intrigued by postcards with images of the Mandelbrot set boundary and of Julia sets. But their main preoccupation and fascination soon would lie with the experiments they were doing.

Armed only with compasses, rulers and maps, some students were discovering that the shoreline of a lake tended to have infinite length as the precision of the compasses increased. Clearly this was due to the detail that a short stride with the compasses picks up but that a long stride jumps over. Since this property persists over many length scales, we learn that at every level of inspection a shoreline resembles itself as a whole.

But such infinite detail is not characteristic of a one-dimensional curve. The shoreline is a fractal, and although its perimeter cannot be defined, the rate at which estimates of the perimeter increase at closer levels of scrutiny is related to its ruggedness. Perhaps we should attribute this to its space-filling ability, or to its complexity ... or both? This characteristic is formalized as the *fractal dimension* of the shoreline.

Meanwhile, another student was breaking rocks with a hammer. When the broken surface of a rock was coated with epoxy glue, a hard smooth shell resulted. When sandpaper was applied to wear down the coating, the rock soon reappeared, as little islands. These islands had fractal coastlines, and the type of rock was identified by the dimension of these coastlines.

Other students watched through a microscope as particles of metal were eaten by acid. Photographs showed how the erosion progressed, and the changes in the metal profile were described in terms of fractal dimension.

Over nine semesters of this pilot study, nine student projects on fractals or chaos won prizes at school or regional science fairs. To these students, to their peers, and to the judges who observed their presentations, fractals were from the real world, and no one wondered if fractals had any applications.

Sometimes dozens of students from seven district schools drove as far as 60 km after school to attend the workshops presented by Dr. Kaye. Often the students were accompanied by their mathematics or science teachers, and sometimes by their English or business teachers. In the same lecture room or lab, these high school students sat among university students who were taking the course for credit, faculty from the departments of geology and astronomy, and visiting engineers from the nickel industry. Everyone was fascinated. Usually, no one was intimidated. Fractals were accessible.

In 1989–90, my school board granted me a sabbatical during which I put together a package of Dr. Kaye's experiments, applications, and conjectures, and some of my own, and continued field testing. I was now certain that high school students would enjoy and benefit from a course on fractals. The details of where it belongs in the high school curriculum would soon have to be solved.

I then offered a one-quarter credit enrichment course at summer school. The students ranged from grade 8 graduates to grade 12 graduates. The interest level can only be described as enthusiastic. Participation in the form of questioning, answering, and conjecturing was something seldom experienced in mainstream mathematics courses. Cries of "Wow!" often could be heard. In the end I made the following assessment:

Fractals and chaos belong in the curriculum!

There is more than enough interesting, valuable material for a complete credit course at the Grade 12 advanced level. Teachers would not have to be asked to pack even more content into already over-stressed courses of study.

My principal, the Board of Education's Director, and the Ontario Ministry of Education approved. So, in the spring semester of 1992, at Nickel District Secondary School, the course, MOS4X Fractals and Chaos, valued at one full high school credit, was taken by a class of ten students as part of their regular timetable. The students' backgrounds were diverse, but one commonality seemed to be their curiosity.

It is somewhat difficult to describe the scope of the course in a single chapter, but let me try to convey some of the main thrusts. After exploring rugged profiles (coastlines, etc.) students are introduced to some mathematically regular fractals such as the Koch snowflake. This curve contains a finite area, but has infinite perimeter, just like the boundary fractals students had been analyzing. We see that this mathematical model has a fractal dimension, calculated by its self-similarity, that agrees with results obtained from perimeter estimates taken with compasses. Next, students construct their own coastline models, by hand, or with a computer program such as *Logo* or *ChaosLab*.

In the third unit of the course, bread and cheese slices are modeled as fractals having no area, but infinite internal perimeter, that is, as *Sierpinski carpets*, with fractal dimensions less than 2. Breads of different textures show different dimensions. As well, Sierpinski carpets can be compared to such things as sections of porous rock taken from oil fields, artificial bone, fibrous filters, and bacterial colonies in a petri dish.

Next, we study the behaviors of long records of coin tosses. Surely, nothing could be so dull. But wait, as you bet a dollar on heads, and I bet a dollar on tails, you are accumulating a considerable lead. Is this fair? The lead has not changed sides nearly as often as you expected. In fact, as we approach the toss you have a lead of about the square root of the number of tosses, and I have not led in the contest for a very long time!

Furthermore, as we examine a graph of the time record of the progress of this experiment, we notice a fractal pattern in the occurrences of the zeros, that is, in the set of times when our wins and losses becomes equal again. This self-similar set of points, called a *Cantor dust*, of dimension between 0 and 1, leads to the notion of fractal time, associated with

earthquakes, avalanches, traffic, and the stretching of polymers.

It is becoming apparent that in nature, fractals are linked to random processes. In several sections we explore random walks and random number tables as we discover some of what Dr. Kaye calls "the surprising patterns of chaos and complexity." Electrolytic trees, viscous fingers, bell curves, and straight lines spring forth. It is now possible to find an answer to the question, "How long is a game of Snakes and Ladders?" We learn how to measure randomness, an elusive concept, as we encounter chaos in the candy store, and other rare events. (Experiments with candy can seldom be repeated.) Then, the ubiquitous π emerges out of the disorder of needles tossed haphazardly on lined paper.

Of course spatial chaos is expected in complex systems such as industrial mixing processes, and social chaos will reign in the absence of the order of law. But what explains the complexity arising from simple equations such as

$$z_{n+1} = z_n^2 + c?$$

In this equation, the z_n on the right side always starts at $z_0 = 0$; z_{n+1} on the left side, the new z_n, is the result of doing the operations of the right side. If at every stage the output z_{n+1}, on the left side, is substituted into the right side, what ultimate result is to be expected? Of course, it is reasonable to expect simple behaviour from such a simple system.

But the answer is that, depending on c, this process, called *iteration*, can lead to any of a large number of possible types of results, including:

- convergence to a *limit point*,

- convergence to an oscillating *cycle* of two limit points, or four, or eight, or . . . ; or three, or six, . . . ; or any number,

- convergence to an infinite set of points, or

- divergence to infinity.

Choose c values from the number line between -2 and $+0.5$. Color code a number with a dull color if the iteration leads to a single fixed point, and use brighter colors for values that lead to cycles of longer period, saving the most brilliant hues for those which shoot off to infinity. You have just created a pixel rainbow, a fractal of intricate beauty.

If z and c are complex numbers, iteration will produce the Mandelbrot set, the points which do not diverge under iteration. And it has a marvellous structure, the most complex and, arguably, the most beautiful of mathematical objects, when viewed in color!

Although the Mandelbrot set is a purely mathematical object, it has fired many imaginations. Philosophers and biologists, among many others, now wonder whether the complexity of life itself arises out of iterations of some such simple formula.

Next, we study strange attractors in weather systems, heartbeats, brain waves, and whale populations, all these modeled by iteration of simple equations. These models, too, exhibit both orderly and chaotic modes. Because the parameters of weather models lead to chaos during iteration, they are so sensitive to error in measurement that long term weather prediction may never be possible. We rejoice in the beauty and despair in the lack of utility of the models for weather. Is nature really chaotic?

While the pursuit of chaos seems to have brought us a long way from fractals, we shall see we remain quite close, indeed. As a parameter is varied, a system passes from one mode to another and from order to chaos and back. There is an underlying structure which students, now near the end of their course, can appreciate. Thanks to a very simple computer program, they can also explore and elaborate and discover for themselves that the pattern in deterministic chaos is self-similar: a fractal!

Associated with this fractal, called the bifurcation diagram, is a new constant of nature. Like π, e, and the golden ratio, the *Feigenbaum number*,

$$F = 4.66920160910299097\ldots$$

is a universal constant of many dynamical systems. In this age of powerful computers, it was discovered with the use of a simple calculator.

In the course, students are slowly enticed to use computers to explore fractals and chaos. Throughout, there have been opportunities to engage computers to produce artificial coastlines, to generate data, or for *Monte Carlo* simulations. Many of the initial programs, written in BASIC or *Logo* are only a few lines long, and seldom are programs more than one page long. Often the annotated program is given in its entirety, and the student is invited to change a parameter, and see what happens. More recently, explorations employ computer spreadsheets, graphing calculators, or interactive programs on the Internet.

Some BASIC graphics programs, including *Iterated Function Systems* and the *Chaos Game*, are simple modifications of a function-graphing routine. And who could resist finding out what the program called *Cascades of Period-Doubling Bifurcations*, can do?

So, we come full circle. Starting with the fractal geometry of nature and on to investigations in chaos we return to order, by way of fractals. What a grand synthesis! What an excellent adventure! What a time to be a mathematics educator!

Since 1992, I have been fortunate to be able to teach my course on fractals in two different schools, a total of three times. In September, 1996, four high schools in the York Region Board of Education, north of Toronto, taught courses on Fractal Geometry and Chaos, based on the one described in this article.

Shirley Dalrymple, a teacher in the York Region Board of Education, started a fractals course in 1995 by submitting a proposal including the rationale for teaching Fractals and Chaos, a description of the existing course that I had prepared, a plan to describe how her team's course would be written in outcomes-based terminology, and a plan to describe how it would be field tested in the three schools. Approval was given for two years. During the first year, a team of three teachers wrote a draft of the course to be field tested and had regular meetings to compare experiences and fine tune strategies. After the superintendent in charge of the project attended a class, questioned the students, and had a first hand look at what it was all about, he advocated strongly for having all schools in the York Board offer the course.

During the second year, the draft of the course was edited and some sections were rewritten. New approaches were tried in classes. Last minute changes were made to reflect the approach of the new provincial curriculum. More of the schools in the York Board are now offering the Fractals course.

Recently the Government of the province of Ontario has initiated extensive reforms of the education system. Most of their initiatives, however, have been related to massive cutbacks in funding and so there is some doubt that resources and morale will be available for maintenance and future development of such initiatives. On the positive side, some opportunity does exist for curriculum revision, which could possibly see fractals topics included, but this will probably depend on the backgrounds of the individuals in charge, and on political considerations. If the positives do not prevail, we will have lost one of the greatest opportunities ever to make mathematics seem relevant and enjoyable for students.

Now we proceed to a more detailed description of some of the experiments in the course.

2 Giant pumpkins and power functions

What do you think is the most popular hobby world-wide? Model railroading, crocheting, a particular sport? No, the answer is gardening, and a growing number of gardeners and non-gardeners alike are becoming fascinated with giant pumpkins.

Figure 1: Giant pumpkins.

Even people who do not actually grow pumpkins themselves travel far and wide in early October to weigh-off sites around the world to observe and especially to touch this huge fruit. Nothing comes close in gardening to the excitement generated by giant pumpkins. Twenty-five years ago, there were tales of a 100 pound pumpkin. But thanks largely to the plant breeding by a Nova Scotian dairy farmer named Howard Dill, the year 1996 witnessed a heavyweight champion, the sumo, the extreme, the ultimate vegetable, a fruit tipping the scales at over 1000 pounds!

The mathematics of pumpkin-growing has become well known to serious growers:

- one leaf produces one pound of fruit;

- fruit weight increases by 30 pounds a day, at peak, in late summer;

- one plant seeded in April will completely cover $200 \text{ m}^2 \approx 2500 \text{ ft}^2$ of garden by late September of the same year.

So many factors bear on the ultimate size of a champion pumpkin that what once was an art has now partly become a science. A seed started in a pot on a warm window sill in the month of May might, by September, after transplanting into a garden, have grown in a single season to produce $450 \text{ kg} \approx 1000 \text{ lb}$ of fruit. If all 450 kg takes the form of a single 450 kg fruit, the grower is looking at a possible world record.

The serious competitor in this field must take great care in cultivating the contender and find every fair advantage. One very helpful thing to know is which pumpkin in your patch is growing the fastest, so that a potential champion can be identified well in advance. Once this is done, all other fruits on that plant are removed to allow all the plant energy to be concentrated in the favourite.

Mathematics comes to the rescue! For the history of modern pumpkin growing coincides with the coming of age of the personal computer. These were combined to the advantage of pumpkin growers by Leonard Stellpflug. He studied graphs of data taken from scores of pumpkins over several years, and used statistical techniques to develop a formula to predict a pumpkin's weight in the field. The presently accepted formula involves the sum of the circumference parallel to the ground, and the ground-to-ground-over-the-top measurement. Following is a table of some results predicted by the formula and used by many serious competitors.

The table can easily be generated on a spreadsheet and individual entries verified on a calculator. The Stellpflug formula for all this is

$$\text{Weight(pounds)} = 0.0000795 \times (\text{inches})^{2.76}$$

where inches means ground-to-ground-over-the-top measurement added to circumference in the horizontal plane.

Horizontal circumference plus ground-to-ground-over-the-top

inches	lbs	inches	lbs	inches	lbs	inches	lbs	inches	lbs
142	69	194	164	246	316	298	536	350	836
144	72	196	169	248	323	300	546	352	849
146	75	198	173	250	330	302	556	354	862
148	78	200	178	252	337	304	566	356	875
150	81	202	183	254	345	306	576	358	889
152	84	204	188	256	352	308	587	360	903
154	87	206	193	258	360	310	598	362	917
156	90	208	198	260	368	312	608	364	931
158	93	210	204	262	376	314	619	366	945
160	96	212	209	264	384	316	630	368	960
162	99	214	215	266	392	318	641	370	974
164	102	216	220	268	400	320	652	372	989
166	106	218	226	270	408	322	664	374	1003
168	110	220	232	272	417	324	675	376	1018
170	114	222	238	274	425	326	687	378	1033
172	118	224	244	276	434	328	698	380	1048
174	121	226	250	278	442	330	710	382	1063
176	125	228	256	280	451	332	722	384	1079
178	129	230	262	282	460	334	734	386	1095
180	133	232	269	284	469	336	746	388	1111
182	137	234	275	286	479	338	759	390	1127
184	141	236	282	288	488	340	771	392	1143
186	146	238	288	290	497	342	784	394	1159
188	150	240	295	292	507	344	797	396	1175
190	155	242	302	294	516	346	810	398	1191
192	159	244	309	296	526	348	823	400	1208

In form, this mathematical equation is called a *power law* because the independent variable, inches, is the base of a power. We are learning that power laws are very common in the scientific description of nature, not at all exclusive to pumpkins. However, we should try to more fully appreciate this law. Consider that if a pumpkin were a solid sphere of uniform density, and if inches stands for its radius, then the equation for its volume is

$$V = \tfrac{4}{3}\pi r^3.$$

If weight were proportional to volume, then adjusting the constant, $\tfrac{4}{3}\pi$, suitably for the units of measurement, we would expect an equation of the form

$$W = kr^3.$$

This differs from the Stellpflug formula

$$\text{Weight(pounds)} = 0.0000795 \times (\text{inches})^{2.76}.$$

Although this power law has nothing to do with fractals (a pumpkin is not filled with holes of all sizes, but only with one large hole; the thickness of the pumpkin wall varies with radius), it draws our attention to the presence of power laws in nature.

3 Is a paper ball three-dimensional?

The mathematician Herbert Robbins once observed that his typical daily activity as a mathematician seemed to amount to filling wastebaskets with paper. He is, however, described by some of his peers as insightful and original in his mathematical endeavours. Perhaps, if he had the benefit of working in the era of fractal mathematics, he would have realized that fascinating mathematics could even be found in a mathematician's wastebasket as Mandelbrot once did when he retrieved a discarded article that turned out to have significance.

But the present problem does not allude to anything written on the paper. Instead it refers to the way that paper crumples.

Before starting the experiment students should draw graphs for the following equations:

- $A = \pi r^2$. Points that come from simple equations often produce simple, but elegant patterns. This curve is a parabola, of course.

- Now, for comparison, we must draw a graph of an equation having an exponent of 3. Let us choose the formula for volume of a sphere: $V = \frac{4}{3}\pi r^3$.

Suppose we use the same r values in exploring both the area and volume equations. Here is a table including some such values. The table gives r values in the left column, while the area and volume function values occupy the two right columns. In between are columns containing values for two other power functions that might be useful for comparison.

Note that the graph of the volume formula points falls on another curve, one that rises even more steeply than the area curve. This is because the sphere is three-dimensional, so the radius measurement was used as a factor three times in the formula.

Now, without changing your data in any way, try thinking of it differently. Think of the circle, not as a circle but as a disk, one unit in thickness. Figure 2 shows the disk from above.

Figure 2: A disk, with concentric annuli, viewed from above.

Think of the area you calculated for the circle, not as area but as mass. So the mass depends on the radius the same way that the area does. Label the first axis on your graph as radius and the second one as mass.

For the sphere, imagine that it is a solid object and think of its volume is its mass. So mass grows faster for an expanding sphere than for an expanding disk when we increase their radii, because the sphere grows in three dimensions while the disk grows in only two dimensions. Note the disk is 2-dimensional and the exponent in its equation is 2, while the sphere is 3-dimensional and its equation has an exponent of 3. Of course.

Figure 3 shows graphs of the equations in the table. We return to them later, for analysis.

r	$y = \sqrt{r}$	$y = 2r$	$y = \pi r^2$	$y = \frac{4}{3}\pi r^3$
1	1	2	3.14	4.19
2	1.41	4	12.6	33.5
3	1.73	6	28.3	113
4	2.00	8	50.3	268
5	2.24	10	78.5	524
6	2.45	12	113	905
7	2.65	14	154	1437
8	2.83	16	201	2145
9	3.00	18	254	3054
10	3.16	20	314	4189
11	3.32	22	380	5575
12	3.46	24	452	7238
13	3.61	26	531	9203
14	3.74	28	616	11494
15	3.87	30	707	14137
16	4.00	32	804	17157
17	4.12	34	908	20580
18	4.24	36	1018	24429
19	4.36	38	1134	28731
20	4.47	40	1257	33510

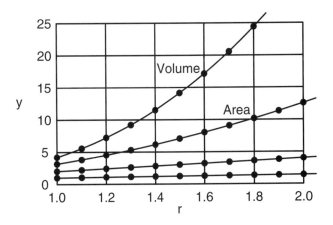

Figure 3: The graphs of the equations from the second table.

Students must notice that on the given scales it is not even easy to tell that, of the four curves, only one is a straight line. These observations are essential. So how do we make them, if we are partially blind to them? We shall see.

3.1 Paper ball experiment

- Take two identical sheets of paper. Crumple one of them up into a tight ball. While holding it tight and round in the palm of one hand, measure its diameter. Plot the *radius you measured* and the *mass of* 1 as a point on the graph.

- Now, tear the other sheet of paper in half. Crumple up one of the halves; keep it tight and measure its diameter. Plot the point with that radius, and 1/2 unit of mass.

- Next tear the remaining half-sheet in half. Crumple up one half of it; keep it tight and measure its diameter. Plot a point with that radius, and 1/4 unit of mass.

- Continue tearing the remaining paper in half, always saving one piece for the next round, and crumpling, measuring and plotting as you go.

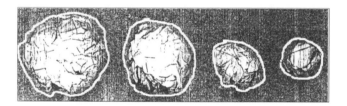

Figure 4: Crumpled paper balls.

Here is a table of typical results.

Radius (cm)	Mass (Sheets)
1.9	1.0
1.4	0.5
1.1	0.25
0.8	0.125

How does the growth rate of this curve, that is, mass of the crumpled paper ball versus its radius, compare to the growths of the circular disk and the sphere? To tell this, plot the logarithm of the mass vs. the logarithm of the radius. This is easiest to do using log-log graph paper. Figure 5 shows the graphs of the four functions of Figure 3, but now transformed by the log-log scales.

Once the curves are straightened out you find that the data for the 3-dimensional sphere falls on a straight line whose slope is 3. So, if the paper balls are 3-dimensional, their graphs will have a slope of 3 too, even if their line is in a different place on the graph. But does it? That is the question. Note the log-log graph of the data for the mass of the

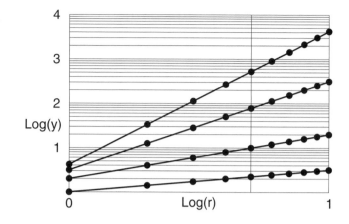

Figure 5: Log-log graph of the data in Figure 3.

circular disk is also a straight line, but its slope is only 2 (as expected). So, what is the slope for the paper ball? An experiment to find out is also a highly suitable science fair project for students.

3.2 Results

On log-log paper we plotted the data from $A = \pi r^2$, from $V = \frac{4}{3}\pi r^3$, and from the crumpled paper experiment. For the first two, the points fall on two straight lines; for the crumpled paper (data from a real experiment), the points were slightly scattered around a straight line.

Now, for the best part: the interpretation of results, and the conclusions. The slope of the first line was 2; the slope of the second line was 3. Students notice

- The circular disk is growing 2-dimensionally, has an exponent of 2 in its equation ($A = \pi r^2$), and produces a line of slope 2 on log-log graph paper.

- The sphere can expand in 3 dimensions, has 3 for an exponent in its equation ($V = \frac{4}{3}\pi r^3$), and produces a line of slope 3 on log-log graph paper.

Therefore it is reasonable to postulate that for the paper balls, if they are in fact 3-dimensional objects, their data line should have a slope of about 3, too. But,

- the slope is only 2.44, not big enough to round up to 3; but certainly too big to round down to 2. So we leave it as 2.44.

This forces us to admit that the paper ball is not 3-dimensional. It is not 2-dimensional—just look at it. So, we concluded that the paper ball (just a crumpled piece of ordinary paper) is a 2.44-dimensional object! And we would therefore not any longer be surprised to find that the data for this object has an equation

$$M = 1.93r^{2.44}.$$

(We also found out where the 1.93 in this equation comes from, but we won't get into that here.) The students had no trouble accepting this conclusion: that the paper ball has a dimension of 2.44. They had suspected early on, that there was too much vacant space inside the paper ball for it to be a 3-dimensional solid like a billiard ball. Even though the paper balls were difficult to compress beyond a certain degree they were not getting solid. Instead they were getting stable. The crumpling was producing an intricate internal structure of wrinkles, and wrinkles on the wrinkles, and . . . . That allowed the balls to resist compression. But except for these structures, the balls were still largely devoid of solid matter.

The students were also quick to see that many other common things are like this: sponges, cheese, and beer foam were suggested right away.

Finally, since the paper ball has a dimension of 2.44, it is called a fractal. While the meaning of fractal continues to evolve, certainly something is a fractal if its dimension is a fraction, not a whole number. Well, up until about 1975, mathematics taught that objects in the real world could only be modelled with integer dimensions. But those who looked carefully saw some subtle problems. For example,

- From far away, a ball of yarn looks like a point, of dimension 0, an integer.

- From closer, the ball starts to look like a circular disk, 2-dimensional, integer again.

- From very close, the ball curves away in all directions, like the earth from space; now the ball of yarn appears 3-dimensional!

The dimension seems to depend . . . on the distance from which the ball of yarn is viewed. So, pre-1975 mathematics mainly left it at that. All dimensions are integers.

But what if we look even more closely at the ball of yarn? The individual strands are not smooth, but covered with mountains and crevices. The roughness persists on even closer inspection. So we must be careful with the notion of dimension for many common objects. Like the crumpled paper ball, many natural things display no awareness that they were supposed to be so confined to integer dimensions.

When Benoit Mandelbrot wrote *The Fractal Geometry of Nature* (Mandelbrot (1982)) he was "merely" observing what my students thought was fairly obvious the day of this experiment: that nature is not constrained to integer dimensions. In fact, so many objects are fractals that Mandelbrot has effectively turned the question around 180 degrees, and we now ask, are there any objects in the universe that are really 3-dimensionally solid?

4   Fractals by electrolysis experiment

This demonstration can be conducted in a classroom in fifteen minutes, with the apparatus set on top of an overhead projec-tor. The experiment can be varied and elaborated in a great many ways, but is presented here in its simplest form. This experiment is a physical realization of a mathematical process called *diffusion-limited aggregation*, or *DLA*, a rich source of fractals. In Section 6 we investigate a computer simulation of this process.

4.1   Purpose

To study the pattern of electrolytically deposited zinc on a carbon electrode.

4.2   Hypothesis

Zinc will be deposited as a *very smooth, circular* disk, layer-upon-layer, around the negative electrode point that touches the solution. This is because the electrolyte solution is uniform in density throughout the petri dish, and the applied electric field is assumed to be symmetrical.

4.3   Apparatus and procedure for electrolysis

a) Cut a narrow strip, 27 cm long, from a sheet of zinc metal.

b) Insert the strip as a lining around the inner perimeter of the inverted cover of a petri dish, diameter 8.5cm.

c) Wrap a piece of fine zinc wire snugly around an end of the zinc strip and connect it to the positive terminal of a 6 volt d.c. power supply.

d) Drill a small hole through the centre of the base of the petri dish and insert a carbon pencil lead so that it just barely extends all the way through. The carbon can be secured with a drop of glue.

e) Connect the carbon electrode with an alligator clip and wire to the negative terminal of the d.c. power supply.

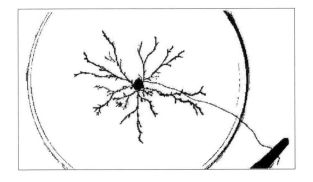

Figure 6: Photograph of a fractal electrolysis experiment.

f) Pour a thin film of saturated, but not over-saturated, zinc sulphate solution into the inverted petri dish cover so that when the base of the petri dish sits on the solution the two surfaces are coated with the zinc sulphate film. The zinc strip is in contact with the film and the carbon point just pierces the film.

g) Close the electrical circuit and observe the voltage with a voltmeter, if available. Expect current of roughly 0.1 ampere.

h) The growing crystal can be photographed at two-minute intervals; alternately, it can be videotaped.

4.4 Theory

Electrolysis: In a solution of zinc sulphate, the potential difference between anode (positive electrode) and cathode (negative electrode) causes the migration of positive zinc ions through the solution to the cathode while negative sulphate ions are repelled by the cathode and attracted to the anode. The zinc ions which contact the cathode are neutralized and stick. On the other hand sulphate ions which reach the zinc anode are neutralized when they capture a zinc molecule from the anode and draw it into solution. The solution is thus replenished and in a condition of dynamic equilibrium.

Kinetic theory of matter: However, even a well-mixed solution of zinc sulphate should not be thought of as a homogeneous medium. The zinc ions have a random arrangement so that we expect submicroscopic clusters near the cathode and elsewhere. This implies that the first zinc particles to be deposited on the cathode might form a fractal (rugged) boundary around the cathode. Both the physical geometry and the electric field of the cathode would be slightly distorted, influencing subsequent deposition. If such a fractal deposition of zinc onto the cathode gets started to even a small degree, we expect it to self-perpetuate.

Coulomb's Law: Within the electrolyte, any zinc ion not very near an electrode will feel much more electrical force from the surrounding ions than from the imposed field. Thus the motions of most ions at any given time is random, according to the kinetic theory of matter.

Shielding: As ions undergoing Brownian random walks approach the cathode, they are shielded from the cathode by branches already deposited. Thus the zinc is more likely to grow more branches than to fill in the region near the cathode.

Sticking: The higher the voltage at the cathode, the higher the probability that a zinc ion will stick to a branch it contacts. Thus the higher the voltage, the less dense the deposited zinc is expected to be.

4.5 Observations, conclusions, questions

1) The current reading was an indication that the zinc sulphate film was conducting electricity.

2) Almost immediately a rather dense growth of zinc is observed on the cathode.

3) Some secondary branches gradually develop on the sides of the primary branches. Little growth takes place at the centre after the branching begins. The initial growth into branches seems to support the theory that at the submicroscopic level the distribution of ions is not uniform.

4) As the branching growth continues, it lends credence to the concept of shielding of the cathode by the branches.

5) The appearance of greater density of growth in experiments where the voltage is held lower seems analogous to the effect produced by lowering sticking probability in the computer model (Section 6). With lower voltage, the ions taking two-dimensional random walks in the thin solution film have a greater chance of avoiding sticking at the tips of branches and are able to wend their way toward the cathode where they might stick to a secondary branch or start a new secondary branch.

6) The zinc tree itself becomes the cathode. Thus the geometry of the tree not only dominates the space within the dish but distorts the electric field as well. In spite of this complicating factor, the DLA tree growths are predicted by the computer model that only accounts for the random motion of ions in solution.

4.6 Applications

DLA structures define a new class of fractal geometric shapes appearing in, for example, electrolysis, rock cracks, and aggregates and agglomerates in fumes and colloids. In a later experiment (Section 7) the dendritic intrusions of a low-viscosity fluid into a higher-viscosity fluid in a Hele-Shaw cell will be studied and compared to DLA forms.

1) This experiment suggests strategies for charging a weak car battery. If a high voltage is used in an attempt to quicken the rate of charge, dendritic structures grow on the cathodes. If these get long enough, they join plates of the cells, causing an internal short circuit (dendritic failure).

   Even dendrites that do not reach this critical length are so delicate they can be fractured by vibrations from the operation of the vehicle. As broken dendrites collect at the bottom of the battery, the build-up can short the battery plates from below (debris failure).

2) Metal Powder Production: DLA is evident in all electrolysis processes, for the deposition of a metal on a cathode is preceded by random walks of ions in the electrolyte and concludes with sticking to the cathode. In industry, rapid electrolysis produces flaky deposits that can be readily scraped from the cathode yielding valuable metal powder. The surface energy of the powder, related to its fractal dimension, and to details of the electrolytic process, determine the rate at which the powder can be sintered.

3) Electronic Circuits: In the electronics industry, small circuit elements are often given a coating of a metal such as gold by the electrolytic deposition of metal vapour. The deposition is a DLA process giving fractal deposits that may not cover the surface densely or continuously enough to make the element electrically conductive. Too rapid deposition may form a delicate structure, resulting in the types of problems described for car batteries.

4) Pigments: Produced by a fuming process, pigments have optical properties determined by their fractal structure.

5) Filters: Dendritic capture trees (dust bunnies) that form on a filter during use, increase the efficiency of the filter.

5 Fractal dimension of electrolytic DLA

5.1 Methods

a) On a photocopy or photograph of the electrolytic DLA draw a series of tightly spaced concentric circles, as shown in the partial diagram below.

b) For each circle, record its radius and the number of intercepts it makes with the fractal tree. Thick parts of the tree intersecting a circle must count as two or more intercepts.

c) Plot a graph of mass (number of intercepts) versus radius, as in the previous paper ball experiment, on log-log graph paper. (Caution: Since it was observed that shielding favours tip growth and inhibits interior growth, the intercept counts at large radii should not be included on the graph, that is, the interior has virtually finished growing, while the tips have not.)

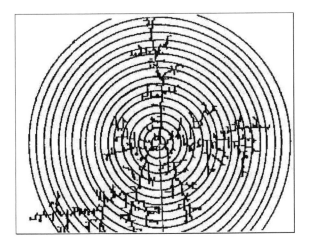

Figure 7: DLA with concentric circles.

d) Compare the results to those of the paper ball experiment.

5.2 Observations, conclusions, questions

1) Is the DLA a fractal object? It certainly appears to be: parts of each branch are similar to branches at other scales.

2) Its mass is distributed as a fractional power of its radius.

3) The tree is rugged and irregular, formed by (chaotic?) processes akin to random walks, while at the same time controlled by physical laws.

What does the DLA have in common with the paper ball? Fractals, or at least power laws, are at work here to be sure.

6 Computer fractalysis experiment

6.1 Purpose

To simulate electrolytic deposition by a Monte Carlo method in which a computer program produces random motion and aggregation.

6.2 Procedure

In the computer model, collisions are assigned a sticking probability, assumed to be analogous to potential difference in the electrolysis application. The annotated program listing (Appendix 1) is quite self-explanatory, and has the following main features:

a) It starts with a circle of 5-unit radius. At the centre is a fixed pixel, called the *seed*, shown in Figure 8.

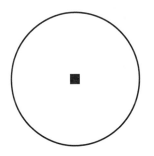

Figure 8: The start of a computer fractalysis experiment: a central seed in a circle.

b) The computer generates a random number between 1 and 360 to select another pixel, this time on the circle. This pixel then begins a random walk. That is, at every step the pixel is equally likely to move one unit in any of the available directions.

c) If the pixel ever contacts the seed, it will stick, with certainty if the sticking probability is chosen by the program user to be 100 %; otherwise it might not.

d) If not, the pixel continues to wander until it might strike the seed again.

e) Later random walkers can stick to any particle already sticking to the seed, as illustrated in Figure 9.

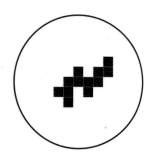

Figure 9: A later view of the fractalysis experiment: several random walkers have stuck to the seed.

Thus a fractal cluster, a simulation of a DLA, is formed. Below is a picture of a screen in which the cluster is composed of scores of particles. A random walker if seen near the top.

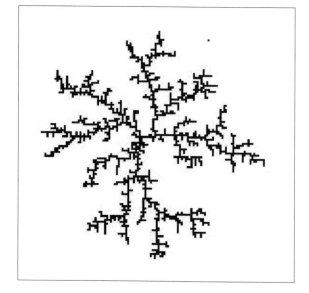

Figure 10: A still later view of the fractalysis experiment.

6.3   Observations, questions, conclusions

In DLA generated by computers, the resulting aggregates are delicate dendritic (tree-like) forms whose density seems related to the ability of the particles to stick on contact. The fractal tree of zinc deposited on the cathode bears close resemblance to the computer simulation of DLA. This supports the model for electrolytic deposition.

1) Shielding refers to the tendency of walkers to stick to outer branches and twigs. Ask your students in what sense is the DLA shielding itself?

2) If the sticking probability is less than 100%, pixels might not stick on contact with the cluster. Ask your students why a low sticking probability would make the DLA denser?

3) DLAs are fractal with respect to their ruggedness and complexity and are more or less self-similar at different levels of inspection. Compare a late-stage computer DLA shown on the left of Figure 11, to the enlargement of its earlier stage.

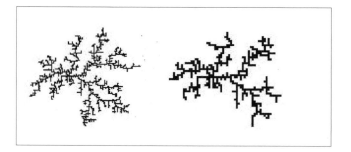

Figure 11: Left: a computer DLA. Right: an enlargement of an earlier stage.

4) Branches of these trees resemble tortuous passages in porous bodies such as oil-bearing sandstone. Understanding the branching structure may have economic significance. A good research project for students is to investigate creative ways that oil is recovered from porous rocks after the underground pressure is no longer sufficient to force the oil to the surface.

5) Another good project is modifying the program to color-code the sequence of arrival of particles at the cluster. This demonstrates the shielding of the inner portions by the outer branches of the tree.

6) The DLA trees are reminiscent of some structures found in the Mandelbrot set, although the reason for the connection is not apparent: there is no randomness in the Mandelbrot set. These Mandelbrot structures, often described as lightning bolts, are actually filaments whereby miniature copies of the set are attached to the main set. See Figure 12.

6.4   Further experiments

a) A simulation of DLA growth can be done manually on graph paper. Use digits 1 to 4, or 1 to 8, of a random number table, depending on how many directions you wish to use. Otherwise the procedure is the same as in the computer program.

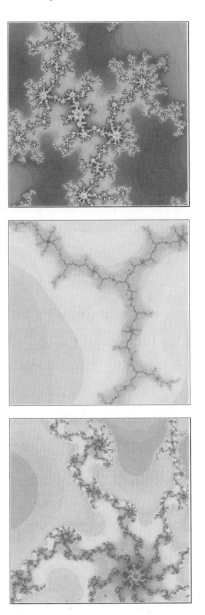

Figure 12: Several magnified views of portions of the Mandelbrot set.

b) Save images of the progress of the electrolytic growth of a DLA. Using image processing software, subtract the first image from the second to observe the screening effect.

c) Experiment with different sticking probabilities, or alter other features.

6.5 Applications

1) filter efficiency

DLA is related to the process by which dendritic agglomerates (also called capture trees) composed of trapped fine particles, grow on fibres in a filter. It appears that the capture trees with high fractal dimension are efficient at catching fine particles, but are delicate and may be broken by the pressure or momentum of the moving fluid and dragged through the filter. Capture trees grown with lower sticking probability are denser, and if more of these dense trees form, the filter will be efficient and strong.

2) carbonblack clusters

An important application of DLA is the formation of clusters of carbonblack particles. Carbonblack is a pollutant in diesel exhaust, but for use in tire manufacture it is purposefully produced by controlled incomplete combustion of natural gas. The fractal structures of these agglomerates determine properties or qualities of the materials they constitute.

In addition, the scientific literature on DLA includes references to these topics.

3) precipitates from colloidal solutions

4) clusters in primordial solar nebulae

5) nerve cells

6) fumes

7) pigments

8) crystals on thin films

9) cracks in metals and rocks

10) breakdown of a dielectric (insulator) due to discharge

11) electrolytic deposition

12) random dendritic growth

13) the dissolution of porous materials such as the etching of powdered aluminum by acid

14) geological dendrites in formations of clay or carbonate sediments

15) igneous fractal patterns where acidic and basic magmas interface

A BASIC program for DLA is listed in Appendix 1.

7 Hele-Shaw cell experiment

7.1 Purpose

This experiment will examine patterns of viscous fingering in a Hele-Shaw cell. Viscous fingering occurs when low-viscosity fluid invades a high-viscosity fluid trapped between closely spaced plates.

7.2 Procedure

A Hele-Shaw cell can be constructed out of the same kind of plastic petri dishes used as electrolytic cells in the experiment of Section 4. As before, the cover of the dish is inverted to become the base of the cell, and the bottom surface of the dish itself is used as the top plate of the cell.

- Drill a 1.5mm hole through the top plate, to match the diameter of a syringe to be used to inject the fluids.

- Squirt some carpenter's white glue into the base of the cell. Make a pool about 2 cm wide.

- Place the top plate of the cell on the glue and press gently to spread the glue to cover about three-quarters of the base area.

- Draw a few millimetres of ink into a syringe.

- Insert the syringe into the hole in the top plate of the cell.

- Gently squirt some ink into the glue.

- Quickly photograph or photocopy the resulting pattern. (It might dissolve.)

See Figure 13. This simulates the injection of water into an oilfield.

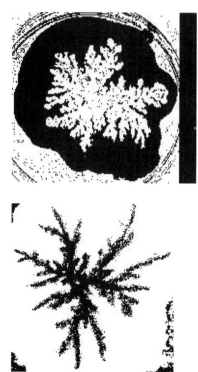

Figure 13: Top: Ink (shown in white) representing water, injected into glue (shown in black) representing oil. Bottom: A color-reversed variant of the top figure.

Observe how some oil is trapped between the fractal fingers of the invading fluid. With practice, and trial of different fluid combinations, a variety of interesting patterns can be produced. Design an experiment that would explore some of the variables.

8 Air fingers experiments

8.1 Procedure

- Prepare a Hele-Shaw cell with glue as before. (Toothpaste could be substituted for glue; two sheets of transparency film for the petri dish.)

- Press the plates together to squeeze the glue, then release the pressure and pry the cell plates slightly apart on one side.

8.2 Observations, questions, conclusions

1) As the glue contracts under surface tension, we see a strange fingering pattern, perhaps resembling an aerial photograph of a landscape. See Figure 14.

2) This time there is no second fluid invading the glue. Or is there? Find out what happens if you apply and release gentle finger pressure on the cell plate directly over any air bubble in the glue. If there are not any air bubbles, try tapping the hole in the top plate to force air into the glue.

Figure 14: An air intrusion experiment.

9 Fractal pancakes experiment

9.1 Procedure

- Heat a griddle or frying pan to about 200° C (400° F).

- Follow any recipe for pancake batter calling for at least one egg.

- Thicken the batter by adding more flour than called for by the recipe.

- Place some cooking oil or margarine on the hot griddle or frying pan and let it warm up for about one minute until it spreads well, covering the cooking surface.

- Pour 50ml of batter onto the griddle.

- Add water to the batter in the mixing bowl until the batter is quite runny. Then pour another 50ml of batter onto a different part of the griddle.

- What forces cause the thick (more viscous) batter to spread out into a disc with a smooth boundary?

- The thin (less viscous) batter expands by fingers along its edge. Is this comparable to the fingering observed in a Hele-Shaw cell, and if so, is there a second fluid with higher viscosity?

9.2 Applications

1) Landfill Sites

We currently lack an appreciation of the movement of buried toxic waste in landfills. Most landfills are lined with a concave bed of clay 1 m to 40 m thick. Even though cracks are sealed with a slurry of clay and soil, migration of toxic wastes through the bed still occurs by diffusion. Our Hele-Shaw cell experiments investigated a fractal fingering model of diffusion.

2) Secondary Oil Recovery

Oil-bearing sandstones are presently our major source of petroleum: the oil is held in the pore spaces of the sandstone. In the initial phase of recovery the oil is driven to the surface at the well site by the pressure of the compressed natural gas lying above the oil. Since 30% to 60% of the total oil in a field can remain in the ground after the primary recovery phase, and since an inexhaustible supply of primary oil fields is not assured, methods to enhance secondary recovery are of great economic importance.

Once the natural underground pressure dies off, a secondary phase of oil recovery begins. One approach is to pump carbon dioxide into the underground chamber above the oil. This pressurized gas drives the oil through the interconnected pores in the sandstone to a second well or group of wells by way of which the oil reaches the surface. As an alternative to carbon dioxide, water is injected below the oil reservoir, but this requires deeper drilling.

Assuming inexhaustible supplies of carbon dioxide and water, two important classes of problems must be solved before maximum recovery can be realized.

i) The first involves the nature of the interaction of fluids. The driving fluids (the carbon dioxide or water mentioned above) are immiscible with the oil; both have lower viscosity than the oil. The differences in viscosities cause interface interaction that take the form of fractal fingering, similar to the DLA shapes we have seen in petri dishes. In the oilfield, the movement of the driving fluids will be along fingers while large pools of oil between the fingers will remain static, at a tremendous economic loss.

ii) The second problem is concerned with the possible fractal nature of the passageways connecting the driving wells to the recovery wells. Although an oil reservoir may be a nearly three-dimensional body, the tortuous transport routes through it may have a lower dimensional fractal form known as a percolation cluster. In Figure 15 a porous rock is simulated with the use of randomly placed squares.

Figure 15: A percolation simulation, with a path, top to bottom, traced.

Notice that the pockets of oil in this model (composed of black squares occupying roughly 60% of the region shown) are embryonic fractals. One percolating pathway is identified with a white line. Even if we learn to eliminate fingering by altering the interaction of the fluids, the pore structure or crack structure of the oil-bearing rock could still impose fractal transport patterns, resulting in the entrapment of vast quantities of oil between passages. It is important to note that the scaling (self-similarity) of fractals suggests that experiments conducted on a small scale in a laboratory will yield results that apply to the oilfield.

As another example of porous substances, think of systems of caves. Cave system structure is self-similar over several orders of magnitude; cave size distribution varies according to a power law. A fractal system of caves is not unlike the supposed structure of porous rock.

It is proposed, furthermore, that patterns of oil seepage may share some geometry with

3) the spread of a plant species in a meadow,

4) the spread of a disease,

5) electrical conductivity of alloys,

6) the lining up of atomic spins to produce a magnet,

7) Artificial Bone

Another area of application of the movement of fluids through porous bodies is in determining the structure of synthetic bone, such as hydroxy apatite. Recently, many people have been able to replace worn or damaged parts of their bodies by prostheses. Bone is a living tissue composed of calcium phosphate crystals embedded in collagen, a fibrous protein matrix. Bone has mechanical properties comparable with cast iron, but is more flexible and four times less dense. Artificial bone is made by heating a mixture of ceramic powder and salt until sintering of the ceramic takes place. The salt is then dissolved and flushed away with water, leaving a porous ceramic solid.

The artificial bone not only is light-weight and strong, but the human body accepts it and adapts to its presence by bonding to it and growing into its pores. Eventually the artificial substance is replaced with natural bone. In terms of a body's acceptance of artificial bone implants, success may prove to be related to the fractal dimension of the bone texture.

In the United Kingdom, in 1970 it was estimated that 5000 hospital beds were occupied constantly by elderly victims of fractured hips, at a cost in 1990 value of about \$200 000 000. Artificial bone is not a remedy for all these problems, but is one method of easing suffering for some, especially those with life expectancies of one or more decades.

8) Tentacles of government

Sometimes the needy are missed by social programs, often due to lack of awareness of available assistance. Could fractal dimension represent the effectiveness of government policies, or the communication of them to the public?

To learn the fundamentals of the complexity and lack of uniformity of the structure of porous bodies, such as those examined or discussed in this article, will require a new geometry. At this time, DLA is the favoured candidate.

10   Do we understand fractal dimensions?

For those of us who have spent our lives up until now confined to a space of three dimensions (all integers), we might wonder whether the fractional (fractal) dimensions of the likes of paper balls and DLAs can indeed be understood, and are, in fact, real. I would challenge the reader to supply definitions of both terms, understood and real. To avoid these philosophical issues, let us look at a simplified way to compute the dimension of self-similar sets:

$$D = \frac{\log(N)}{\log\left(\frac{1}{r}\right)}.$$

The D in the equation is the similarity dimension. Suppose that we magnify a fractal object, say, to find the crinkle on a wrinkle of crumpled paper, or a branch of a branch of a DLA. Knowing the magnification, we determine the size of, say, a DLA branchlet relative to the branch it is part of: that relative size is r in the equation. To look at it another way, $\frac{1}{r}$ is the magnification that brings a branchlet to branch size. Denote by N the number of branchlets that make up a branch, assuming that a branch is made up entirely of branchlets. Probably the most famous example of an object that can illustrate the use of the equation is the Koch snowflake curve of Figure 16.

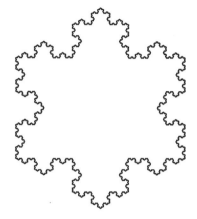

Figure 16: The Koch snowflake curve.

Figure 17 shows a coarser version of the snowflake in which some of the textural details have been shaved off.

In this diagram, a portion of its coastline was magnified, as indicated by the radiating lines. The magnification factor was 3, that is, $\frac{1}{r} = 3$. The magnification produced 4 times as much detail as before; so $N = 4$. Therefore, the fractal dimension of the Koch coastline is,

$$D = \frac{\log(4)}{\log(3)},$$

approximately 1.26. This number implies that the coastline is more rugged than a smooth line, but being less than 2-dimensional, cannot actually cover an area. Nevertheless, seen to the detail of Figure 16, it does look fuzzy. That the fuzziness does not quite amount to surface area is emphasized on magnification, when the border always reveals a profound emptiness. However, being self-similar it always retains enough fuzziness for another magnification.

The formula $D = \frac{\log(N)}{\log(\frac{1}{r})}$ can be used not only to analyze, but to create fractal images as well. Suppose, for example, an object is found experimentally to have a dimension of 1.7.

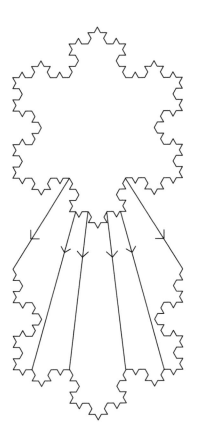

Figure 17: A coarsened snowflake, with a portion magnified.

A DLA produced by electrolysis is such an object. But if we have very little experience with fractal dimensions, then we might wish to gain a better appreciation of what a dimension of 1.7 means. One way to start is to compare the object visually to a collection of mathematically constructed objects having a range of dimensions. We can use the dimension formula to construct a mathematical fractal having a given dimension. All we have to do is to pick a dimension, D, and then find an N and an r that will satisfy the equation, fit the pieces together to cover the initial stage, rescale, replicate, repeat, ... The determination of N and r can be done by trial and error, in which case a computer is very useful, or with a little algebra and trigonometry. Appendix 2 shows a range of fractals of different dimensions, produced by three different generators (motifs), at various stages. Try to find the profile of a familiar face in this gallery.

As this chapter comes to a close, let us glimpse some figures produced by the Chaos Game, so often used to introduce potential fractaliers to the Sierpinski Gasket. To play the Chaos Game, specify a set of vertices in the plane, and a scaling ratio r, a number between 0 and 1. Select a starting point, and choose one of the vertices randomly. Draw the line from the starting point to the chosen vertex, and place a point on the line between the starting point and the vertex,

r of the distance to the vertex, $1 - r$ of the distance to the starting point. Repeat the process, with the new point taking the place of the starting point. Ignore or erase the lines. The collection of points rapidly fills in an image, often a fractal. Figure 18 illustrates the Chaos Game to make the Sierpinski gasket. Here the vertices are those of an equilateral triangle, and the scaling ratio is $1/2$.

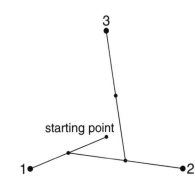

Figure 18: The Chaos Game with vertices of an equilateral triangle, and scaling ratio $1/2$.

Figure 19 shows a collection of fractal dusts and doilies was generated by varying the rules of the Chaos Game slightly, with special attention to the scaling ratio, r, and the replicating factor, N (the number of vertices), so as to produce objects of the desired dimension D.

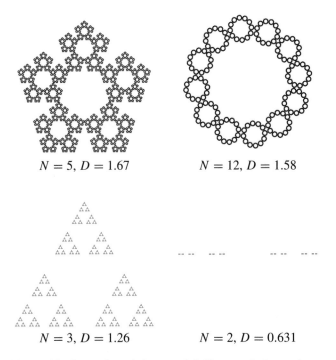

$N = 5, D = 1.67$ $N = 12, D = 1.58$

$N = 3, D = 1.26$ $N = 2, D = 0.631$

Figure 19: Some fractal dusts and doilies, made by variants of the Chaos Game.

The following observations may be of interest to the pure mathematician or the numerologist.

- The first figure, with five vertices, employs the Golden Ratio. With this ratio the five main images which compose this fractal seem to just touch. The fractal dimension of the figure is close to that of DLA.

- The second figure has twelve vertices, but has the same fractal dimension as the Sierpinski Gasket, widely discussed in other works.

- The third figure is not actually the Sierpinski Gasket because the pieces do not meet. They are dust. But this figure has the same fractal dimension as the Koch snowflake border.

- The fourth figure, with only two vertices, is the dust of the Cantor middle thirds set.

The student interested in generating mathematical fractals to mimic nature's shapes would do well to start with the Chaos Game, especially its variant the *Iterated Function System*, or *IFS*, for short. Besides considering fractal dimension, more realistic imitations of nature also would have to account for the property of lacunarity (roughly, how uniformly the holes are distributed—a topic for another day).

Our demonstration of ability to generate and analyze fractal models should begin to convince even the skeptic that this notion of fractal dimension is both understandable and real. Furthermore, there is a whole world of fractal applications, indeed, a fractal universe waiting to be explored. What a time to be a mathematics student!

Appendix 1

Computer Program (BASIC): DIFFUSION-LIMITED AGGREGATION

```
DIM freq(1500), radint(1500)              'allocating memory for statistics on
                                          'distribution of particles
Setup:                                    'Randomwalker initiates from inner
                                          'circle, which is always 2 pixels from the
                                          'nearest point of the cluster. Walker is
                                          'annihilated if it wanders beyond the
                                          'outer circle.
CLS                                       'clear the screen
INPUT "Sticking probability (in %; try 100)"; prob
                                          'used in Stik
INPUT "Pixel size (magnification; try 10) ="; s
n = 1: freq(0) = 1: radius(0) = 0         'start counting with seed
p = 245: q = 145                          ' seed at screen centre
IF s =3D 1 THEN
  PSET(p, q), 33                          ' 33 is a color
  ELSE
  LINE (p - s/2, q - s/2) - STEP(s, s), 33, bf

                                          'seed scaled by s

END IF
rad = 2*s                                 'radius of inner circle from
                                          'which migrating points
                                          ' begin their randomwalk

Begin:
  'CIRCLE (245, 145), rad , 3, , , .89     ' inner circle 'color 3
  'CIRCLE (245, 145), rad + 2*s, 2, , , .89  'outer circle 'color 2
Randompoint:
  RANDOMIZE TIMER                         'computer clock used to seed
                                          'random number generator

x = s*INT(RND*(2*rad/s + 1)) - rad        '*Students should investigate
y = (2*INT(RND*2) - 1)*(rad - ABS(x))     'how these equations give
                                          'the coordinates of point of a
                                          '"taxicab," or "pixel space" circle
p = 245 + x                               'place diagram at mid-screen
q = 145 - y
Flash:
  IF s =3D 1 THEN
    PSET(p, q), 33
    ELSE
    LINE (p - s/2, q - s/2) - STEP(s, s), 33, bf
                                          'draw randomwalker
  END IF
  GOTO Stroll
CheckNeighbours:
  IF s = 1 THEN
    lefton = POINT(p - 1, q): righton = POINT(p + 1, q)
    topon = POINT(p, q - 1): boton = POINT(p, q + 1)
    ELSE
    lefton = POINT(p - s/2, q): righton = POINT(p + s/2, q)
    topon =POINT(p, q - s/2): boton = POINT(p, q + s/2)
END IF
```

```
        IF lefton = 33 OR righton = 33 OR topon = 33 OR boton = 33 THEN
                                        'check whether any neighbouring
                                        'pixels are occupied
            GOTO Stik
            ELSE
            GOTO Rub
        END IF
    Stik:                               'walker contacts cluster
        c = c + 1                       'count all contacts
        IF INT(RND*100) + 1 < = prob THEN   'to stick or not to stick
            n = n+1                     'count sticking contacts
            radius = SQR(x*x + y*y)     'record position of contact
            radint(n) = CINT(radius)    'nearest integer radius
            freq(radint(n)) = freq(radint(n)) + 1   'updated frequency at present radius
        Flash2:                         'draw new pixel sticking
            IF s = 1 THEN
        PSET(p, q), 33
        ELSE
        LINE (p - s/2, q - s/2) - STEP(s, s), 33, bf
                                        'draw randomwalker
            END IF
            GOTO NewCirc
        END IF
    Slip:                               'contact without sticking
        IF righton = 33 THEN
        righton = 1
        ELSE
        righton = 0
        END IF
        IF boton = 33 THEN
        boton = 1
        ELSE
        boton = 0
        END IF
        IF lefton = 33 THEN
        lefton = 1
        ELSE
        lefton = 0
        END IF
        IF topon = 33 THEN
        topon = 1
        ELSE
        topon = 0
        END IF

    neighb = 1*righton + 2*boton + 4*lefton + 8*topon
                                        'check which neighbouring
                                        'pixels are occupied
    ON neighb GOTO DLU, RLU, LU, RDU, DU, RU, U, RDL, DL, RL, L, RD, D, R
                                        'no step will be retraced
                                        'nor crossed
        DLU: ON INT(RND*3) + 1 GOTO Down, Left, Up
                                        'possible moves depend on
                                        'neighbouring points
```

```
        RLU: ON INT(RND*3) + 1 GOTO Right, Left, Up
        LU: ON INT(RND*2) + 1 GOTO Left, Up
        RDU: ON INT(RND*3) + 1 GOTO Right, Down, Up
        DU: ON INT(RND*2) + 1 GOTO Down, Up
        RU: ON INT(RND*2) + 1 GOTO Right, Up
        U: GOTO Up
        RDL: ON INT(RND*3) + 1 GOTO Right, Down, Left
        DL: ON INT(RND*2) + 1 GOTO Down, Left
        RL: ON INT(RND*2) + 1 GOTO Right, Left
        L: GOTO Left
        RD: ON INT(RND*2) + 1 GOTO Right, Down
        D: GOTO Down
        R: GOTO Right
Rub;                                    'erase pixel
    IF s = 1 THEN
    PSET(p, q), 33
    PSET(p, q), 30
      ELSE
      Flash3:
    LINE (p - s/2, q - s/2) - STEP(s, s), 33, bf
                                        'glimpse at position
    LINE (p - s/2, q - s/2) - STEP(s, s), 30, bf
                                        'erase current walker to background
                                        'colour, i.e., last position turned off
      END IF
Bounds:
    x = p - 245: y = 145 - q
    IF ABS(x) + ABS(y) > rad + 2*s GOTO Begin
                                        'should point wander too far
                                        'annihilate it and start another
Stroll:
    IF s = 1 THEN
      PSET(p, q), 33
      PSET(p, q), 30
      ELSE
    Flash4:
      LINE (p - s/2, q - s/2) - STEP(s, s), 33, bf
                                        'glimpse at position
      LINE (p - s/2, q - s/2) - STEP(s, s), 30, bf
                                        'erase current walker
    END IF
    dir = INT(RND*4) + 1                'choose one of four random directions
    ON dir GOTO Left, Down, Right, Up   'move point in chosen direction
      Left: p = p - s: GOTO CheckNeighbours
      Down: q = q + s: GOTO CheckNeighbours
      Right: p = p + s: GOTO CheckNeighbours
      Up: q = q - s: GOTO CheckNeighbours
NewCirc:
    x = p - 245: y = 145 - q
    IF ABS(x) + ABS(y) < rad - 2*s GOTO Begin
                                        'if cluster is at least three
                                        'units from inner circle
```

```
        'CIRCLE(245, 145), rad , 0, , , .89              'erase current circles to be
        'CIRCLE(245, 145), rad + 2*s, 0, , , .89         'replaced by larger ones
        rad = rad + s
        IF rad > 179 THEN STOP                           'circle is larger than screen
      GOTO Begin
      END
```

Appendix 2

A Gallery of fractals with specified dimensions

Dimension	Generator	Stage 2	Stage 3	Later Stage

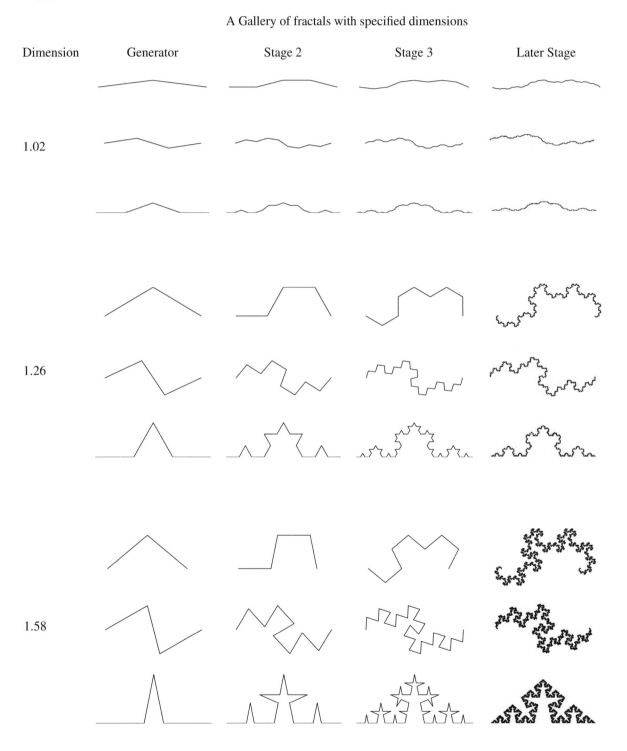

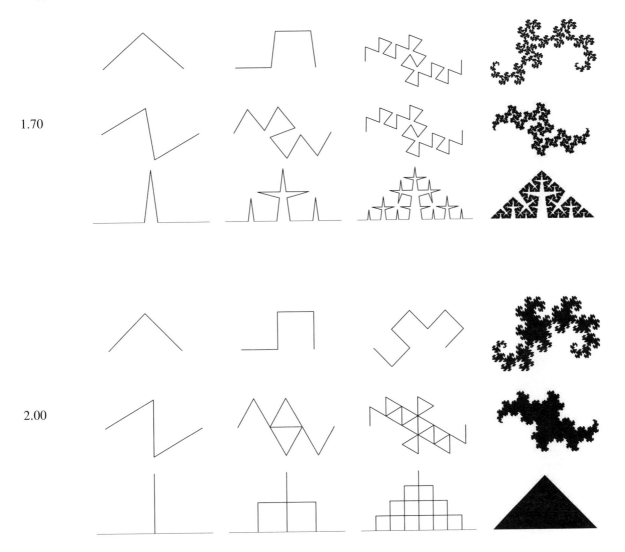

Chapter 13

Fractal Themes at Every Level

Kenneth G. Monks

OK, I admit it. I love fractals. Fractal programs, fractal T-shirts, fractal notebooks, fractal screen savers... What other mathematical phenomenon has made it into mass marketing? I have had students walk into my classroom carrying folders with fractal designs on them, unaware that they were fractals. This is a phenomenon we rarely encounter in mathematics, and one on which we should capitalize.

Teaching fractals and chaos often necessitates the use of special texts, software, hardware, and other instructional materials. In a small institution on a limited budget, an instructor often is forced to improvise or make due with the equipment at hand. However, with a little bit of planning and forethought a math department can make some wise decisions that can make the necessary tools available. One purpose of this chapter is to review the instructional materials I have used, and to describe some of the software I have used or produced for my students use in these courses.

I have been incorporating fractal geometry and chaos theory into my teaching at the University of Scranton (a Jesuit undergraduate liberal arts university in northeastern Pennsylvania) to varying degrees since 1990. The degree has ranged from minor bonus projects in linear algebra and complex variables to entire courses, including a freshmen level general education course for non-technical liberal arts students, and a junior level course for students who have had the second calculus course. The teaching of fractals has also touched a wide range of audiences outside of my normal courses, ranging from a presentation in front of a group of second graders, to mentoring award-winning undergraduate research projects in mathematics, to teaching two undergraduate courses at Yale. Like fractals themselves, the teaching of fractals seems to have something of beauty and value to impart at all levels.

1 Which courses can benefit?

1.1 Second grade elementary school students

Without a doubt, the most enthusiastic audience I have ever talked to about fractals was my daughter's second grade class.

I have yet to see a group of college students leaning on the edge of their seats and waving their hands wildly ("Ooh, ooh, call on meeee!!!") as these students did in the hope of being allowed to answer a question about math! I had been invited to give a talk on mathematics to this wonderful group of kids and decided that an introduction to fractal geometry would be a perfect topic.

I brought with me the university's LCD panel which allows me to project the images from my notebook computer using a standard overhead projection unit. The night before I had loaded many fractal images (generated mostly by FRACTINT (see Section 3.2)) in a directory on the notebook for easy recall. After a brief fractal slide show (which they received quite enthusiastically), I began the main topic of the talk, namely an introduction to the Sierpinski Triangle (a.k.a. gasket).

There are many age-appropriate activities involved with this idea. We began with a large triangle on the blackboard with the children taking turns subdividing it in the usual manner (we put a safety dot in the center triangle formed after each subdivision to protect it from further subdivision). This involves the concept of finding the midpoint of line segments in increasingly complicated figures, the concept of a line segment connecting two points, and the idea of a recursive geometric construction. At each stage of the construction, we also counted the number of new triangles which did not contain safety dots, and made conjectures as the construction progressed about how many non-safety dot triangles would be formed at each stage (a nice illustration of a geometric sequence and pattern recognition).

After this I showed them the Chaos Game (using the students themselves as points on the floor and dice to choose between the tree vertices) and returned to the computer to show how the Chaos Game generates the Sierpinski triangle.

At the end of the presentation, I handed out large paper triangles and straight edges. They then constructed paper Sierpinski triangles of their own. They are not very competent with rulers at that age, but many more than half of the students were able to make a quite respectable looking Sierpinski triangle.

1.2   Freshman general education course

In the Fall 1992 and Spring 1993 terms I taught a sequence of two special topics courses designed to meet the science general education requirements for freshmen liberal arts majors. General education requirements were under extensive review at the time and I felt that the traditional topics covered in a course of this nature needed to be reconsidered.

I had recently finished producing the color plates and some other figures for Davis (1993). Don Davis was my thesis advisor at Lehigh University, and although our area of research is algebraic topology, while I was still a graduate student we were bitten by the bug to learn more about the wonders of fractal geometry. I had known about fractals from the standpoint of a computer hobbyist long before becoming a mathematician, but my first actual mathematical course in fractal geometry was in a once per week seminar on iterated function systems taught by Don at Lehigh during the Spring of 1989. As a result, I naturally decided to teach the freshman courses out of photocopies of Don's book (which had not yet been published at the time).

The first course was on fractal geometry and number theory. In it we covered roughly Chapters 3 and 5 of Davis (1993), as well as some supplementary material of my own. In order to provide a more unifying theme to the course, I tried to center it around the concept of recursion in both geometry and number theory. I used the connection between Sierpinski's triangle in fractal geometry and Pascal's triangle in number theory (see, for example, Section 8.3 of Peitgen, Jurgens, & Saupe (1992a)) as the link to make a smooth transition from the fractal geometry section of the course to the number theory section of the course. This approach seemed to work very well and gave the course a unified feel.

The theme of the second course was non-Euclidean geometry, and concomitantly a lot of time was spent on formal axiom systems and the nature of mathematical proof. This course covered roughly Chapters 1 and 2 in Davis (1993).

As it turned out, the first course was a far greater success than the second in terms of student enthusiasm. As a mathematician, I feel that the material covered in the second course is one of the most beneficial topics we can teach to liberal arts majors for their general education exposure to mathematics. It reveals the very nature of mathematical thought, and illustrates the profound impact that the non-Euclidean revolution has had on our perception of mathematical truth. Yet, the students did not seem to share my enthusiasm for the topic.

I attribute this to the fact that it is simply more enjoyable and accessible for the typical liberal arts student to make color blow-ups of the Mandelbrot set or construct Pascal's triangle than it is to work on proofs or compare the axioms of non-Euclidean geometries with those of Euclidean geome-

try. The fractal geometry and number theory students had a good deal of enthusiasm for the material throughout the course while I got the sense that the non-Euclidean geometry students viewed it as a chore, something to be mastered even though they had no interest in it.

Looking back, I should have known this would be so. It takes several years before even our math majors have a decent idea of the concept of a mathematical proof. To expect that liberal arts freshmen would be able to obtain a deep enough understanding in only a few short weeks to appreciate the difference between the geometries is now, in my opinion, unrealistic. That is not to say that they did not benefit by struggling with the material. I feel that they did, but I also feel that my course goal of imparting to them a general understanding of the impact of the non-Euclidean revolution on the nature of mathematical thought was certainly not attained, and perhaps is unattainable with that particular audience in the limited time available.

With fractal geometry and number theory, however, one can teach the material at many different levels, starting from the level of the second grade presentation if necessary, and gradually adding rigor. Students can produce zooms of the Mandelbrot set, for example, even before they know its definition or any of its properties. This keeps them connected to the material in a fundamental way when you *do* proceed to discuss the definition and properties. It is the cover design that entices the students to open the book. With the non-Euclidean geometry course, it is much more difficult to find a suitable carrot.

The software requirements of the freshman fractal course were very minimal and are discussed below. There were no software requirements for the non-Euclidean geometry course.

1.3   Chaos course

During the Spring of 1994 and again in the Fall of 1995, I taught a three credit junior level course entitled *Chaos and Fractals* geared toward students who had at least a Calculus II background and a basic familiarity with computers. I circulated an advertising flyer of sorts at registration time and there was a lot of student interest in the course.

The class was primarily composed of math and computer science majors, but I also had chemists, a physical therapy student (the physical therapist changed his major to mathematics after taking the course!), an economics major, and a philosophy major who audited. In addition, one of my colleagues in the math department sat in on the 1994 course.

The course met three days per week throughout the term. My original intention for the course was that it would be mostly about chaos theory, using Devaney (1992) as the primary text (required) and Peitgen, Jurgens, & Saupe (1992a) as a secondary text (optional, but strongly recommended to the students). Both books contain an excellent description of the topics of chaos theory, but the lack of exercises in the latter precluded its use as a primary textbook. Devaney's book does not go into nearly as much depth in fractal geometry, so it was necessary to have the secondary book available to the students.

The original plan was to cover nearly all of Devaney's book, pausing every second Tuesday to interject some topic from fractal geometry. By the end of the course, the connections between these fractal digressions and our study of dynamical systems became apparent.

After the course was completed, several students mentioned that they had signed up for the course hoping to learn more about fractals than they did. They were disappointed at the low fractal-to-chaos ratio of the course. I must agree. There is not much you can accomplish in only six or seven lectures on fractal geometry. As a result, when I offered the course again in the Fall 1995 term, I devoted a much larger percentage of the course to fractal geometry, covering topics from fractal geometry once per week on what was affectionately dubbed Fractal Fridays. To compensate for the extra fractal material in the course, I covered the first five chapters of Devaney's text (which discuss iteration, orbits, graphical analysis, and periodic points) at a much more rapid pace, as most students had no difficulty in this part of the course the first time it was taught.

The course went quite well once again. But the students had two suggestions for improvement at the end of the course. Just as with the first course, they had hoped there would be a larger fractal geometry content to the course rather than the roughly 2/3 chaos to 1/3 fractals. Secondly, they felt that it would be more coherent to do all the fractal geometry lectures together and all of the chaos lectures in another contiguous block. Often from one Fractal Friday to the next they would forget definitions and topics discussed the week before, making it hard to follow the theme of the fractal lectures.

As a result of the overwhelming success of this course, it has now been added as a permanent listing in our catalog, to be offered once every two years. The prerequisite for the course is now one mathematics course beyond Calculus II. The decision to strengthen the prerequisite was prompted mainly because of the difficulties that the students had understanding the proofs in the course, so they could better appreciate the theory as well as the applications and results.

1.4 Fractals and chaos at Yale

During the Spring 1998 semester I had the honor of being invited by Benoit Mandelbrot to teach two courses on fractals and chaos at Yale University. It was a very rewarding experience to have worked with Mandelbrot, the father of modern fractal geometry. My conversations and interaction with this legendary innovator and genius have expanded and altered my perspective on fractal geometry, and all of mathematics,

in many ways that will impact my teaching of the subject in the future. I am very grateful to have had this opportunity.

The two courses I taught at Yale were geared towards completely different audiences. One was a freshman level course designed for liberal arts majors (similar to the course I taught at Scranton using Davis's book). The second was a course designed for sophomore-junior level students who have had calculus (similar to the course I taught at Scranton using Devaney's book). In the rest of this chapter I will refer to these two different types of courses as the *liberal arts course* and the *junior level course*.

In the liberal arts course at Yale, I used the excellent textbook Peak & Frame (1994). We were able to cover nearly all the material in this book in a single semester. The course focused on the major topics in fractal geometry and chaos theory from a mathematical standpoint, but in addition had many illustrations and references to applications in other areas of interest to liberal arts students. Students were required to produce a term project relating fractal geometry and chaos theory to their own interests, and some of these projects were quite well done and interesting (fractal patterns in dance, music, art, literature, chaos in baseball statistics, and CD player track changers to name just a few).

In the junior level course I decided to used Crownover (1995) as this book has a much higher fractal to chaos content ratio and covers the topics of fractals and chaos in contiguous blocks of lectures rather than trying to intersperse them with each other as I had done in the past. I also had Mandelbrot (1982) and Peitgen, Jurgens, & Saupe (1992a) as recommended texts. The choice and level of topics covered in Crownover is very good for a course like this. However, the exposition leaves much to be desired and within a few weeks both the students and I had become so disenchanted with the text that we stopped using it altogether, instead I prepared my lectures using material from Peitgen, Jurgens & Saupe (1992a) and Barnsley (1988).

In the end, the course focused on the study of iterated functions systems and their mathematical applications, and the basic theory of chaos. The students were also required to produce a term project, most of which were quite impressive. Both courses went quite well and student feedback was overwhelmingly positive.

1.5 Fractals in other courses

In addition to the courses discussed above, many traditional courses can benefit from a healthy dose of fractal geometry.

In linear algebra, I assign a project on iterated function systems at the point in the course where we study linear and affine transformations of the plane. The project consists of a handout which explains the iterated function system concept, and then leads the students through the computation of a Sierpinski Carpet. The students are then asked to produce a Sierpinski Triangle via iterated function systems. The project does not usually generate the amount of enthusiasm that I would expect, perhaps because the students feel it is just additional work on top of the main topics of linear algebra many of them are struggling to learn.

Naturally, in the complex analysis course I cannot resist giving the definition and a short project on the Mandelbrot set. This serves as an illustration of complex arithmetic, the geometry of subsets of **C**, and the complicated action of an iterated quadratic polynomial as a mapping of the complex plane. This is also given as an outside assignment. As in linear algebra, almost no lecture time is taken away from the primary subject matter.

In general, I feel that these projects did not add (or detract) substantially to the courses for most students, and I now believe that the correct place for such material is in special topics courses devoted to them such as those discussed above.

1.6   Undergraduate research projects

In 1991 the University of Scranton instituted a Faculty Student Research program. In this program, students receive transcript recognition for no fee or credit upon the approval of the mentor. At the start of each term in which the student participates, the mentor and student sign a contract of sorts, stating what will be required of the student during that term. If the contract is met, then the student gets transcript recognition stating that he participated in the program, otherwise nothing happens.

Most of the student projects to date involved topics related chaos and fractals: Fraboni (submitted), Farruggia, Lawrence, & Waterhouse (1996), Fusaro (1997), and Joseph (submitted). All of these undergraduate student papers would make interesting supplementary reading in an undergraduate chaos and fractals course and would provide the students with an example of the kind of mathematics research that an undergraduate is capable of doing. Nearly all of my student researchers were former students from my chaos and fractals course, indicating the mathematical enthusiasm that this course is capable of generating. For a more detailed description of the content of these papers, see the papers themselves or Monks (preprint). The papers themselves are available at

http://facweb.scranton.edu/~monks/

fractalthemes.html

2   Textbooks

In this section I give brief informal reviews of the textbooks I have used in my fractals courses.

2.1   The nature and power of mathematics

As mentioned above, I have taught a two-course sequence using Davis (1993), covering most of the material in it. I must

also warn the reader that I might not be the most objective reviewer of this book; first, because the author, Don Davis, is my Ph.D. thesis advisor and friend, and second, because I produced all of the color plates and many of the black-and-white diagrams for the book, and was a major proofreader of the original drafts.

That having been said, I must say that the book is a very well written exposition, and does a masterful job of explaining extremely difficult mathematical concepts, at a very low level, while maintaining the mathematical integrity of the subject matter. This is not an easy balance to achieve. Yet Davis succeeds in doing this extremely well. As he says in his introduction:

> This book was written for the liberal arts student, but it could also be read with benefit by a good high school student, ... or by an educated general reader who is willing to do a little work while reading.

The book succeeds in this respect as well.

The exercises range from very basic problems to some that are far too difficult for a normal liberal arts student. This is not a standard text in the sense of having a lot of worked examples for the students to imitate when attempting the homeworks, but rather the students are expected to work the exercises in order to develop a deeper understanding of the concepts introduced in the text.

In addition to the material on fractal geometry, the major topics covered in the text are formal axiom systems, non-Euclidean geometry, and number theory and cryptography. Each of these can provide excellent supplementary material in a course designed around the fractal geometry core material. Alternatively, the instructor may wish to supplement the fractal material in the book with additional topics if a fractals-only course is desired.

2.2   First course in chaotic dynamical systems

Devaney (1992) is a perfect text for teaching an undergraduate course on chaos. The book can be read easily by the students, the exercises are just the right level of difficulty, and the book comes with an excellent set of laboratory experiments intended to be assigned as student projects. I used several of these projects with great success.

The pacing of the book is a bit slow in the first five chapters (which discuss iteration, orbits, graphical analysis, and periodic points), and I recommend that you cover this material rather quickly, as my students had very little difficulty with it. The meat of the course is contained in Chapters 6 through 12 (bifurcations, quadratic maps, symbolic dynamics, chaotic maps, Sarkovskii's theorem, etc.). This material should be covered carefully by the instructor, as students do have difficulty with the amount of analysis and topology required to go through this material. This is unavoidable, but

Devaney does an excellent job of teaching only those concepts from these areas that the students require to understand the key concepts and results of chaos. The book is certainly appropriate for students who only have a Calculus II background.

If no outside topics are introduced, one should easily be able to cover the entire book in a three credit course. If a substantial amount of fractal geometry from other sources or other material is to be covered early on in the course, then the material in Chapters 13 through 16 on Julia and Mandelbrot sets and Newton's method fractals can be omitted without much loss, as these topics would have already been covered.

This book is almost entirely a book on chaos, so if you want to teach a course that thoroughly covers the key ideas in fractal geometry, you might want to use a different text. For chaos theory and discrete dynamical systems, Devaney's book is excellent.

2.3   Chaos and fractals

Peitgen, Jurgens, & Saupe (1992a) is a true work of art in every sense. It is organized in such a way that it covers the basic ideas for a non-technical reader, while interjecting mathematical rigor for the more sophisticated reader in a very unobtrusive way. In addition to covering many topics from fractal geometry, the book also presents very well the material of chaos covered in Devaney's text. The illustrations are simply fantastic.

This might make an excellent text for a course on fractal geometry and chaos theory and provide the instructor with a lot of leeway for selecting topics (as there is far too much material to be covered in a three credit course). It has been either a recommended or required text in nearly all of my chaos and fractals courses to date.

The only serious drawback this book has as a text is the lack of exercises. I strongly hope that someone will produce a companion exercise manual for this text in the future, or that exercises will be incorporated into future editions. Even still, the extra work required to make up problem sets is somewhat compensated for by the sheer joy of being able to cover such a wide variety of topics. It is also the kind of book many students might think twice about selling back to the bookstore at the end of the term.

2.4   Chaos under control

Peak & Frame (1994) is the perfect choice for a freshman level liberal arts course. It is written at a level that assumes almost no mathematics on the part of the reader and yet gently nudges the student into an understanding of very difficult mathematical topics without sacrificing mathematical rigor. The authors go to great pains to describe the mathematics accurately, but with a minimum of mathematical jargon.

The material is constantly illustrated with applications and implications to areas outside of mathematics that might be of interest to the typical liberal arts student. The book is very up-to-date, and discusses many of the recent developments in the areas under discussion. It covers the major topics from fractal geometry and chaos theory, and also a lot of applications of both to the study of randomness and data analysis (something not found in most of the other texts on chaos and fractals). This gives the course a decidedly applied flavor which appeals to the liberal arts students without compromising the mathematical integrity of the material. The pace of the book is perfect and I was able to cover almost the entire book in a single course which met three hours per week for one semester. I used this book again when I taught the course at Scranton.

The only minor flaw in the book is a lack of included exercises at the end of each chapter (there are a few in the back of the book). But this was not a problem as one of the authors, Michael Frame, has a complete exercise set available at his web site.

```
http://classes.yale.edu/math190a/CUC/
              welcome.html
```

Additional material can be found at

```
http://classes.yale.edu/math190a/Fractals/
              welcome.html
```

The author was also very helpful to me during the course via email and provided other instructional materials such as solutions to the homework sets, some nice fractal animations, lecture notes, and so on. With the easy availability of the exercise sets via the internet, there is no reason not to use this text in a freshman liberal arts course on fractals and chaos, or even in a freshmen general education course designed to give students the flavor of modern mathematics and mathematical research. I think the topics are more exciting and relevant than some of the other traditional texts that cover more classical mathematics, and Peak and Frame's book worked well for our general education math course at Scranton. I give their text my highest recommendation for this kind of course.

2.5   Fractals and chaos

Crownover (1995) has all the markings of what should be an ideal textbook for a sophomore-junior level undergraduate text on fractal geometry and chaos. It has an excellent choice and order of topics, covering fractal geometry first and then leading into several extensive chapters on chaos theory. It has slightly more material on fractal geometry than chaos theory, but the topics and order chosen blend well into a cohesive syllabus for such a course. It appears at first glance to be written at the right level of difficulty for students, and contains many exercises. All of this led me to select it as the primary textbook for the advanced undergraduate course at Yale.

However, only a few weeks into the course, both the students and I decided to abandon the book in frustration. The exposition is very uneven and confusing in general. The students found the material hard to read. I had difficulty preparing lectures from this text, not because the presentation is too advanced for such a course, but rather because of the lack of care in defining terms and symbols and proving theorems, and because of the mathematical errors beyond simple typos. I ended up correcting and revising the text material so much in my lectures that it became easier to simply abandon the text altogether and have the students learn everything from my lectures.

The author seems to be making an honest and sincere effort to create the perfect textbook, and I have no doubt that in a future edition it might be an excellent text to use in such a course. However, much revision is required before that can be attained. In its current form, I cannot recommend this text to others. Of course, it is possible that the author and I simply have different perspectives on how the material should be presented, and thus you might want to check out the text to decide for yourself.

2.6   Other texts

No discussion of textbooks on fractals and chaos theory would be complete without mentioning the seminal treatise, Mandelbrot (1982). This book is the *de facto* recommended reading in my courses as it is the original source of inspiration from which the subject has developed. It both provides historical background for the subject and is a wonderful source of applications and ideas to inspire student projects, especially in an advanced undergraduate course.

The book is written as a classic reference work documenting the ideas and discoveries of Mandelbrot, not in the expository style of a textbook intended for classroom use. It is very dense, each chapter being the equivalent of one or more traditional research articles. Thus the book is not well suited for use as a primary textbook in an undergraduate course. It is too difficult for most students, who would find the densely packed collection of abstract ideas and concepts very difficult going without the careful detailed hand-holding that exemplifies a typical textbook. However, as a reference work and source of inspiration, it is a true classic that I highly encourage my students to read, and often refer to as a source of applications and historical development when developing my lectures. I highly recommend it to any instructor in a course on fractals and their applications.

During my recent advanced undergraduate course at Yale, I had an opportunity to prepare many of my lectures using Barnsley (1988). While I have not used his text as the primary required textbook in any course thus far, I was very impressed with both the depth and variety of the topics covered, the con-

sistency and rigor of the mathematical exposition, and the truly excellent collection of wonderful exercises at the end of each section. The book is in some sense complementary to Devaney's in that Barnsley's book is almost entirely about fractal geometry with very little chaos discussed instead of the other way around. I was very impressed and I would consider using this as a primary textbook in the future in a course which was primarily about fractal geometry for undergraduates with a calculus prerequisite.

3   Software requirements

Finding or developing the appropriate software for courses on fractal geometry or chaos requires a bit of planning. In a small undergraduate institution it is a good idea to incorporate software and hardware planning into a larger picture that considers the department's needs as a whole.

For this reason our department has centered our computational needs around *Maple*. The flexibility of *Maple* makes it possible to use it in almost all of our undergraduate courses that require (or can be enhanced by) the use of a computer. We currently have a mathematics computer lab housing nineteen Pentium II PCs running *Maple V R5* under *Windows 95*. In addition, *Maple* is available on some public labs and also under VMS (text only) on our academic mainframe and on other WIN95 machines around campus. However, students primarily use it in our math lab setting.

3.1   Maple

Maple is flexible enough to satisfy the needs of most courses and provides the students with a uniform tool that they can use as they move through our curriculum. In my Chaos course, in the beginning of the term I make available a *Maple* notebook Monks (w) containing many of the routines the students would require to complete their assignments (I did not want to spend a lot of time teaching *Maple* syntax).

The routines consist of the usual array of tools for studying chaos and fractals. There are routines to compute orbits, plot graphical analysis diagrams, make bifurcation (orbit) diagrams, and illustrate the chaos game. They are listed in the notebook with examples showing their use.

The orbit routine comes in two flavors: exact and approximate. The approximate orbit gives the values of the orbit floating point form with its concomitant roundoff error, while the exact version uses *Maple's* exact arithmetic, which is quite useful for studying periodic points.

Also available is a routine that produces an animation of the graphical analysis diagram so that students can actually see the construction of the diagram in real time, rather than just the finished product.

I decided to provide these routines to the students in the form of a *Maple* notebook rather than putting them in the

Maple library. This way, the students have access to the source code that they can modify or learn from when creating *Maple* routines for their projects.

All of the projects assigned from Devaney's book were completed on *Maple*. A particularly interesting and difficult project is the determination of Feigenbaum's constant (Experiment 10.4 of Devaney (1992)). This is a very tricky calculation because of the proximity of the roots to each other. Students learned very quickly that one cannot always rely on *Maple*'s built-in *fsolve* routine to magically come up with the answer they desire. There were many different approaches to this problem, some good, and some not so good. I used *Maple* to come up with a solution using a simple bisection method for finding the roots of and using the graph to determine an interval containing the desired root.

However, despite its sophistication, even *Maple* has some limitations. For example, *Maple* is somewhat slow in producing even the crudest plots of the Mandelbrot or Julia sets. It can be helpful in analyzing the fate of individual points, but its graphics capabilities and slow computation speed limit its ability to produce fractals.

However, some of these limitations have been overcome recently with the introduction of hardware floating arrays in Release 5 of *Maple*. By doing floating point computations directly in the host's hardware, the user can avoid the slow arbitrary precision floating point routines of *Maple*. Recently I discovered the excellent *Maple* fractal software at

```
http://www.math.utsa.edu/mirrors/maple/
frame03.htm
```

that overcomes many of these limitations and I will certainly use this software and modify my own routines to incorporate hardware floats for use in my future courses.

3.2 FRACTINT

This is the mother of all fractal programs, and what's best, it's free! This DOS program (there is a Windows version, but the DOS version is more robust) is a public domain fractal generation program that is very fast and very complete. I used version 19.6, but FRACTINT is being revised constantly to add new features. I have not encountered any program in its league for doing Mandelbrot set zooms. It is blazingly fast and runs on a variety of video hardware to produce stunning pictures in the highest resolutions your system supports. Almost every conceivable fractal type is supported. I have had students in my courses use it to produce Mandelbrot and Julia type fractals, IFS fractals, and to generate L-systems. It is also available by anonymous ftp on internet at many different sites (e.g. Monks (w)). We always keep the latest version installed on all relevant machines on campus, since it costs us nothing and benefits the students. I have walked into a computer lab to find students who know nothing about fractals

and who are not even in a course requiring its use, playing with the program. This program is a must for every math department.

3.3 Calculators

Currently, we do not have available an easy method for displaying computer graphics in the classroom. The classroom I usually teach in is equipped with a computer and overhead projection system that is time consuming to hook up and activate (see the discussion of projection units in Section 4.1). Also when doing the homework problems, it is often overkill to be chained to a PC running *Maple* and many students in our school still do not have a personal computer. A solution to some of these problems comes in the form of the programmable graphics calculator. Several students purchased one of these calculators for use in my chaos course.

When I taught the chaos courses in 1994 and 1995, I used a TI-85 programmable calculator. The TI-85 is an affordable and sophisticated programmable graphics calculator. In Monks (w) you will find a scaled down version of the *Maple* utility discussed above written for the TI-85 that I wrote for my own personal use in the course (doing homework, making test problems, answering spontaneous questions that arise in the classroom). It basically is a menu driven program that allows the calculator to compute orbits and draw graphical iteration diagrams.

One nice feature of the TI-85 calculator is the ability to transfer programs from one calculator to another via a cable that comes with the calculator. Several of the students in my class also purchased the TI-85 and I gave them this chaos program. In most cases, the calculator is sufficiently powerful to work some of the projects in Devaney's book, and I think that it would be possible in theory (though less desirable) to run an entire chaos and fractals course using only this calculator as a computing platform (although tiny black and white fractals produced on a calculator after literally days of computation are not exactly awe inspiring).

I have written a bifurcation (orbit) diagram program for the TI-85 (Monks (w)). This program allows the user to get numerical information by moving the TI-85's cursor on the finished drawing. Many other programs, including ones to draw the Mandelbrot set, are available for the TI-85 on the internet.

Since teaching the 1994 and 1995 courses I have purchased a TI-92 calculator. This is more expensive (\$165–\$200), but is *very* powerful. Essentially the entire computer algebra system *Derive* is built in, providing arbitrary precision arithmetic and symbolic computation capability, with a standard alphanumeric keyboard.

Thus I did not have to write *any* software in order to use the TI-92 in a chaos course because it already has the capability to compute orbits (using the Table feature, set the Mode to Sequence) and it also has the ability to do graphical analysis built-in (from the y= screen, set `Axes` to `Web`). This calculator is ready to use out of the box in a chaos course!

It is an easy-to-use calculator with more power than rightly ought to be in a calculator! The only bad feature of both the TI-92 and the TI-85 is the poor display. They are *extremely* difficult to read in incandescent lighting. I have to prop my calculator at a certain angle and position all of the lights in the room just right in order to be able to use it for any length of time. It is really an unbearable feature on otherwise fine pieces of equipment and I hope that TI will fix this problem in future models.

3.4   Other software

There are probably hundreds of fractal and chaos programs currently available (see (Monks (w)). Most deal with a specific set of examples or are companion software to certain textbooks. I do not think that anything beyond *Maple* and *Fractint* is required for serious study of fractals and chaos at an undergraduate level. Since *Maple* does double duty by providing superb CAS support to other courses in our department, it is a very cost effective approach to satisfying the computational needs of the department as a whole (as opposed to buying individual programs that service only one particular course).

4   Hardware requirements

4.1   Classroom displays

The classroom that Yale provided for the liberal arts course was outstanding and this influenced the way I taught the course. The room was a medium sized auditorium holding about 100 students with a sloped seating arrangement. There was a ceiling mounted projector that could display video on a screen lowered above the blackboard. On the wall was a control box containing a VCR, audio system, and a video port into which I could plug my laptop. The screen could be lowered to just above the blackboard, and the lighting was on a dimmer switch, allowing the simultaneous use of the blackboard and computer display.

This was an ideal setup for such a course, and as a result I found myself making the course a lot more multimedia intensive. My course directory grew during the term into a collection of 35 different programs, graphics, animations, and Power Point slide shows. Of course, this took a lot of work to develop the first time, but I now have these available for teaching the course in the future.

I cannot emphasize how much such a multimedia approach enhanced the course and aided these non-technical students in learning the material, and I cannot imagine teaching the course exclusively at the blackboard. The ability to discuss the theory at the blackboard and illustrate the results on the video display, simultaneously, was very effective at getting across otherwise difficult and abstract mathematical ideas.

As a result I am now in the process of working with the administration at Scranton to upgrade the computer display capabilities of the mathematics classrooms, so that other courses might benefit from this multimedia approach.

We currently have an overhead projector with a color VGA LCD display connected to a Pentium computer in two of our math classrooms. But I rarely use these because of the time involved in setting them up and the fact that the projection cart and screen obstruct some students' view of the blackboard, making it impossible to use both at the same time. If you are trying to outfit your classrooms with computer displays, be sure to outfit the rooms with the ability to control lighting (blackout curtains to eliminate daytime light pollution, and a dimmer to allow the taking of notes while viewing the projection). Also you should install the computer so it does not require extensive setup and tear-down time. Finally, it is desirable to be able to use the blackboard at the same time as the computer without darkening the room.

I also recommend the use of a laptop for each instructor over the use of a dedicated in class PC. It simply takes too much time to prepare multimedia lectures and get them working on your own computer at home or in the office and then have to transfer the files to the classroom PC (which may not be the same classroom each term) and test it to be sure the required software is installed, there are no version incompatibilities, the hardware difference doesn't cause a problem, etc., etc., etc. With a laptop the instructor can prepare his or her presentation before class and use that same machine for the presentation, eliminating all the transfer and debugging time and assuring the presentation will work as expected.

4.2   Lab computers

As mentioned above we have nineteen Pentium II machines in a very nice mathematics lab. The lab is not limited to computer use, but is rather intended to be a generic student study area for mathematics of every kind. We have a small library of math books, and a discussion area with white boards for the students to talk about mathematics. This room is located in close proximity to the math department faculty offices, making the faculty very accessible to the students working in the lab.

4.3   Programmable calculators

As mentioned above, I have used the TI-85 and currently use the TI-92. These are both wonderful calculators that are capable of streamlining most numerical processes and do a very nice job of graphing (for a calculator). Many students in my numerical analysis course had purchased one of these and seem to be very pleased. It would be nice to have every student own one, but we have not required this in any course so far.

There is also a computer link available that works with either calculator, allowing you to write your calculator programs on a PC and then transfer them to the calculator. The software supplied is elementary and the link is very expensive (almost as much as the calculator itself), but it is absolutely necessary if you want to write programs or use those you find on the internet. You can enter programs from the tiny calculator keyboard, but it is very awkward to do so (especially on the TI-85), making the link necessary for programmers.

5 Conclusions

Fractal geometry and chaos theory have the magical ability to attract students to mathematics, to engage their attention on the material, and to keep it there while teaching them some very difficult and beautiful mathematics. I am certainly not advocating teaching a subject simply because it pleases the students. What makes students happy and what is best for their mathematical education is certainly not always the same thing. However, a happy interested student is much more likely to devote the effort and time needed to master difficult material than is an alienated one. If fractal geometry and chaos can make students *eager* to learn about metric spaces and topological conjugacy, dense subsets, and fractal dimension, then I am all for it.

For a one semester liberal arts course, I highly recommend Peak & Frame (1994). If your university has a two semester liberal arts general education math sequence, I recommend the Davis (1993) for the second semester as it has excellent exercises and a wide variety of wonderful topics beyond fractal geometry and chaos. For a junior level course, there is no ideal textbook that covers both fractals and chaos. Devaney (1992) is an excellent text for a course that emphasizes chaos theory and discrete dynamics but does little fractal geometry. Barnsley (1988) emphasizes fractal geometry iterated function systems. Peitgen, Jurgens, & Saupe (1992a) covers both fractals and chaos very well, but lacks exercises. Crownover's book (1995) has the right mix of topics, but the exposition is lacking.

However, as a result of my experiences, I no longer believe that the topics of chaos and fractals should be taught in a single advanced undergraduate course. There is simply too much important material in both areas to fit both topics into a single course. Instead, I am currently suggesting to our department that we break up the chaos and fractals course at the University of Scranton into two courses. The first course would cover the topics of chaos in discrete dynamical systems that appear in Devaney, and in addition would cover continuous dynamical systems and chaos in differential equations instead of discussing fractal geometry. The second course would be on fractal geometry, and would cover most of the material in Barnsley's book including applications of iterated function system fractals to fractal data interpolation. So it is my hope that our existing course will split into the two courses, one on fractals and one on dynamical systems. (The liberal arts course can still cover both topics in one semester, because it is designed as a survey course anyway.) In all of these courses, the classic text Mandelbrot (1982) should be recommended reading and referred to frequently by both students and faculty.

A program such as *Maple* can satisfy almost all of the software needs of such courses, and does double duty by providing software support in most cases across the mathematics curriculum. This is an efficient strategy both economically and pedagogically. Programmable calculators provide the means for students to have substantial computing power available at the cost of a good textbook, and can be a nice complement to the power of *Maple*, but their use is not essential and need not be required in any of these courses if *Maple* is available to the students on campus.

The topics of chaos and fractal geometry are very timely. It is one of the rare undergraduate mathematics subjects dealing with theorems and definitions that are recent and constantly being refined. During the 1994 chaos course we had been discussing Devaney's definition of chaos, namely that a dynamical system is chaotic if:

1. it has dense periodic points,

2. it is transitive, and

3. it has sensitive dependence on initial conditions.

We then discussed the recent article Banks, Brooks, Cairns, Davis, & Stacey (1992) in which they show that conditions #1 and #2 imply condition #3 for infinite sets. As the course went on, I received the current issue of the *Monthly*, containing Vellekoop & Berglund (1994) in which they prove that condition #2 implies #1 for functions defined on intervals. I then presented these new results in the 1995 chaos course. Since then Touhey (1997) has given yet another definition of chaos, namely that a continuous function on a set is chaotic if every two disjoint open subsets share a periodic orbit.

It is very exciting to be able to watch the definitions of the topic evolve and develop before our eyes, sometimes changing from the start of the course until it is finished. In my chaos courses, we discussed other articles that included simple-to-state open problems in these areas. The opportunity for undergraduate students to experience mathematics as a fresh, living art, in which there are more questions than answers, in which definitions and theories are refined and developed over a long period of time by a group of real living people, is a rare one indeed . . . and one which can do a lot of good for our subject and students.

Chapter 14

Art and Fractals: Artistic Explorations of Natural Self-Similarity

Brianna Murratti and Michael Frame

1 Interactions between art and mathematics

The familiar interaction of art and mathematics involves applying mathematical tools to seek patterns in artistic works, for example, the harmonic analysis of paintings into golden rectangles of Ghyka (1977). This idea is not new: in painting it dates back at least to da Vinci, in architecture to the Parthenon.

The fractal geometry of Mandelbrot (1982) is a profound extension of traditional geometrical tools, and is the first effective language for understanding the roughness of nature. Because many aspects of art are inspired by nature, fractal geometry provides a new collection of tools for uncovering patterns in art. Voss & Wyatt (1993) used multifractal analysis to study early Chinese landscape paintings. Taylor, Micolich, & Jones (1999) demonstrated a correlation between the fractal dimension of Jackson Pollock's drip paintings and the stage in Pollock's career when the paintings were done.

Another instance of mathematics influencing art is the use of computers to produce art, as described in Mandelbrot (1989), for example. Instances include Harold Cohen's artificial painter, Aaron (McCorduk (1991)), and Ken Musgrave's artificial landscapes (Musgrave & Mandelbrot (1991)). Musgrave and others have argued that although a large part of their work involves designing algorithms that to some degree mimic natural processes, the artistic expression is manifested through features including the general form of the landscape and the color palette. As proof of the effectiveness of these landscape forgeries, we mention that after Musgrave presented a slide show in *Fractal Geometry for Non-Science Students* at Yale in 1993, some students insisted one of the slides

was a fake forgery—a slide of a real landscape inserted to see if anyone would notice. Further discussion revealed the students disagreed about which slide was real (in fact, all were forgeries), and afterward some reported they were profoundly disturbed by this ability to so convincingly synthesize natural-looking scenes. One lamented, "I won't be able to trust what I see on television or in the movies." This was several months before the opening of the movie *Jurassic Park*.

A powerful argument for the influence of mathematics on art was made by Shearer (1992) when she put forward the thesis that the major revolutions in art have been catalyzed by revolutions in geometry. The development of geometric perspective influenced Renaissance art (Edgerton (1992)), and non-Euclidean and higher-dimensional geometry had a demonstrable impact on the beginnings of cubism and modern art (Henderson (1983)). Shearer goes on to speculate that fractal geometry may signal another revolution in art. In Shearer (1995) she argues persuasively that geometric thinking is a tool of fundamental importance for artists, and for scientists.

Yet another approach, a precursor to the direction discussed here, is the use of art to enhance students' understanding of science. Peak & Frame (1994) presented a method, based on some of the paintings of Ellsworth Kelly (Kelly (1992)), for visualizing patterns in chaotic processes. These Kelly paintings proved a valuable tool for helping students grasp some of the more subtle aspects of chaos.

This experience led us to speculate that art could be used for communicating some of the more difficult concepts of fractal geometry. Here we report on exercises and demonstrations to explore the differences between mathematical and natural fractals, and to understand the characteristics of chaos, topics that often have challenged our students.

2 Some problems with teaching fractals and chaos to non-science students

Ten years' experience teaching fractals to non-science students at Union College, Yale University, the University of Richmond, and Utah State University have revealed a variety of difficulties. Some of these involved routine pedagogy—developing interactive computer graphics labs and incorporating writing assignments, for instance—but others have resisted the usual approaches. The examples we wish to address here fall into four categories: (a) the limiting form of mathematical fractals, (b) the distinctions between mathematical and natural fractals, (c) the characteristics of chaos, and (d) appropriate choices of material for artistic term projects.

Fractals are characterized by self-similarity or self-affinity. They are composed of pieces looking like shrunken copies of the original shape (with different shrinking factors in different directions, in self-affine cases). In mathematical descriptions, this decomposition sets in motion an infinite regress: the pieces are themselves composed of smaller pieces looking like shrunken copies of the original, and so on. The traditional view of mathematics and nature, grounded in centuries of the spare abstraction of Euclidean geometry, is that nature is noisy mathematics. Mountains are correctly viewed as weathered cones, clouds as crumpled spheres. Nature is studied by first recognizing the underlying Euclidean shapes, and then roughening them. This indoctrination starts early. Children who begin drawing before kindergarten often try to draw trees, mountains, and clouds as they look, within the limits imposed by their skills. In school drawing classes, these realistic attempts are replaced with more abstract, simpler lollipops, triangles, and arcs of circles. A glance out the classroom window will convince any child the drawings aren't right, and yet these Euclidean caricatures are presented as the right way to draw. Small wonder that Mandelbrot's inversion of the traditional relation of geometry and nature met with such opposition.

One of the main messages of fractal geometry is the inappropriateness of this Euclidean approach. Nature is understood by its roughness, viewed through self-similarity. Recall Mandelbrot's formulation, "Clouds are not spheres, mountains are not cones, coastlines are not circles, and bark is not smooth, nor does lightning travel in a straight line." (page 1 of Mandelbrot (1982)) Yet we teach fractals following the Euclidean pattern, mathematical fractals first because (we think) they are easier to understand, then natural fractals as noisy mathematical fractals. (But note that Ron Lewis (Section 12.1) begins his course with natural fractals and later moves to mathematical fractals.) This lack of attention to a main lesson of our field is responsible for many of the problems we have in teaching fractals. We propose inverting this order. First engage students in seeing natural fractals, understanding self-similarity as approximate and applied to a limited range of scales. Then present mathematical fractals as an abstraction of natural fractals. Starting with mathematical fractals causes many problems.

For instance, while limiting processes are familiar to anyone who has mastered calculus, most of our students do not have this background, or have only a rudimentary exposure to it. Indeed, one of our main reasons for developing a course in fractals for non-science students was to offer to these students an invitation to science through the contemporary, rapidly-growing, visually-oriented fields of fractals and chaos. Consequently, when we begin by drawing the first few stages of a mathematical fractal on the board, then say something like "continue in this way and the fractal arises as the limit," no amount of supporting computer graphics has gotten this point across to the entire class. Some persist in thinking the final drawing presented, perhaps the third stage of the construction, is the fractal. Moreover, among those who do understand mathematical fractals, the transition to natural fractals is problematic. Having grasped the limiting processes and self-similarity on infinitely small scales, they have some difficulty replacing these notions with approximate, statistical, self-similarity over a limited range of scales. (See Figures 1 and 2.) The frustration is understandable: the geometrical language developed specifically to study nature seems least effective when applied to natural objects. While many do understand mathematical fractals and are excited by their ability to find very simple descriptions of complicated-looking shapes, they see less clearly that fractals are a geometry embracing the roughness of nature.

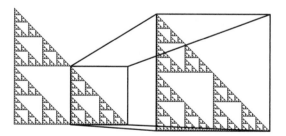

Figure 1: Exact self-similarity on all scales is epitomized by the Sierpinski gasket. Magnifying a square around any part of the gasket reveals a fragment containing an exact copy of the whole gasket. This magnification can be repeated infinitely many times, obtaining smaller and smaller copies of the gasket.

The introduction to chaos follows a parallel path. First we present, usually with some motivation, a simple cartoon model of a chaotic process. The tent map and logistic map are frequent choices. Following Devaney (1989), chaos is characterized by three properties: sensitive dependence on initial

Figure 2: A natural fractal, a dendritic mineral deposit. Magnification would show that the side branches of the dendrite are smaller dendrites, statistically similar to the whole. Further magnification would show that this side-branching continues for only a few levels.

3 Using art to teach fractals and chaos

We now believe natural fractals should be among the first examples presented. In some previous years we have begun by cutting up cauliflower or looking at Queen Anne's Lace. These examples have the relevant properties of natural fractals—approximate self-similarity over a limited range of scales—yet showing them in class did not seem to improve significantly some students' grasp of the ideas. To be sure, these examples, especially the cauliflower, worked for many students. In general, however, they were not as successful as we had expected. What was the problem, and how can we solve it?

Most obviously, a class demonstration does not engage everyone, regardless of how many questions one asks the students. Also, in presenting the demonstration, we establish language describing the relevant points. Students can copy the language and reproduce it without real understanding, illustrated powerfully to us by some peculiar attempts to apply it in other situations.

Probably the clearest example of misinterpreting the demonstrations was an answer to the final exam question, "Describe the differences between natural fractals and mathematical fractals." Several students replied, "Natural fractals live in three dimensions, mathematical fractals live in two dimensions." While unluckily it was true that year the only mathematical fractals we showed were subsets of the plane, this is not the point we wanted students to take from the demonstration.

To approach these problems, we have developed a set of lab exercises in which students are exposed to the properties of natural fractals by creating their own examples. Certainly, for engaging students, individual exercises are more effective than demonstrations. In the lab write-up, students must develop their own verbalizations of these properties. Our intention is for each student's exercise to be different from the others, so although the fractal features will be similar in the large, detailed descriptions will be individual.

The lab exercises use art to create natural fractals. Iterative processing of watercolors, inks, oil paints, finger paints, and several viscous fluids give students a physical appreciation for how repetition acts to produce fractals. Careful examination of their product illustrates the approximate self-similarity and limited range of scales of natural fractals, the analysis and lab writeup give students ample opportunity to put the ideas in their own language. Pre-lab discussion equips students with enough of the vocabulary of art to analyze their projects, and because students work in different media, the post-lab discussion provides a setting for students to present their findings to one another, each discussing the other's work.

For interpreting the characteristics of chaos, mixing is nicely illustrated by allowing watercolors to bleed into one another. Sensitive dependence on initial conditions also is easy: dropping acrylic ink into two watercolor washes, pre-

conditions, mixing behavior, and denseness of periodic orbits. The first two of these are easy to grasp for both the cartoon maps and natural chaotic processes. On the other hand, the third property, central to any study of the important topic of controlling chaos, seems much more difficult.

Finally, in the early stages of development any new field grows wildly in many directions. Especially in applications to other fields, some advances are profound, others are nonsense. For teachers, this is a great, though fleeting, opportunity. For a short time, we can show the human side of science, the contingent nature of progress, the mistakes, the surprises, the genuinely unexpected connections that celebrate the mind's ability to unpack the complexity of the world. Recognizing mistakes is important, too, and not only for pointing out the human nature of science. Skepticism is an essential intellectual skill. In a culture filled with an increasing barrage of pseudoscientific nonsense, we should embrace every opportunity to help our students develop their sense of skepticism. In some ways, this course reflects the same opportunity. We have found term projects to be an important part of the course, and all too often students have executed their projects in ways missing direct contact with the topics of the course. They seem to think for a moment, perceive a connection between a field in which they are interested and some aspect of the course, and build a project around this first idea. While giving students the opportunity to take some of the course material into their own world, the term project also is an opportunity to exercise skepticism, applied to the students' own ideas, and to the most rapidly growing source of information: the web.

Several years of experimentation with these problems have revealed art as an effective approach. In the next sections we shall illustrate some of our observations.

pared as nearly identically as possible, will inevitably flow in different patterns. This is a convincing, almost obvious, example, (reminding some students of Jeff Goldblum's demonstration of chaos with water drops on the back of Laura Dern's hand in *Jurassic Park*) and yet it has a subtle side that is an essential precursor for the next exercise. As applied to chaos, sensitivity to initial conditions refers to behavior in time. Small changes in the initial conditions lead to an eventual uncoupling of the two signals; they grow to differ, at times by an amount the same size as either signal. This exercise freezes the temporal divergence into a difference of spatial patterns; spatial structure becomes a metaphor for temporal behavior. Recognizing this is the key for interpreting some of the viscous fluid experiments as illustrations of denseness of periodic behavior.

Finally, these experiments give a sound foundation for those students wishing to do a term project on some artistic aspect of fractals.

4   Exercises and demonstrations

Easy examples of natural fractals can be made with watercolors. The materials needed are water, a paintbrush, paper (good quality, fine grain watercolor paper, Fabriano, for example), watercolors (those from a tube allow more control of viscosity), and a container for mixing the watercolor paint. To avoid a gravity-driven bleed, place the paper on a flat surface. First, with a mix of $\frac{1}{4}$ water and $\frac{3}{4}$ watercolor of a lighter color (e.g., yellow), paint a thick line. Before this dries, adjoining but with little overlap, paint another line of a contrasting color (red is a good choice). By applying the second color before the first has dried, inevitably the two colors will bleed together. These regions of bleed, where the newer, less-viscous paint is diffusing into more-viscous, partially dried older paint, are natural fractals. Two examples of such a wet-on-wet experiment are shown in Figure 3. Note the branches off branches off branches along the interface. Careful study under magnification reveals the approximate self-similarity and limited range of scales of these natural fractals. Experimenting with the concentration of the paints, how dry the first is before applying the second, how much of the second is applied, and so on, gives students the opportunity to explore variations in an open-ended fashion. Contributing something of their own to the project increases students' sense of ownership of the ideas.

A slightly more complicated preparation gives a natural fractal pattern closely resembling the mineral deposit of Figure 2. The materials needed are water, a paintbrush, finger paint paper, watercolors (again, those from a tube are preferred), acrylic ink (permanent), a dropper, and a container for mixing the watercolor paint. Start by wetting the side of the finger paint paper to be used. (The correct side to use is shinier and smoother.) Mix watercolor from the tube with

Figure 3: A watercolor bleed, growing a natural fractal.

water in a 1-to-1 ratio. Paint the watercolor onto the damp paper. While the watercolor is still rather wet (let it set no more than 20 to 30 seconds), drop one drop of the acrylic ink onto the paper. The results are immediate - the ink flows into the watercolor, diffusing in a filamentary pattern, branches off branches. See the top of Figure 4.

Figure 4: Top: Acrylic ink on a watercolor wash. Bottom: A dendritic finger paint fractal, formed as two painted surfaces are pulled apart. Several different spatial frequency components are indicated.

With an opaque projector or an overhead mounted video camera and projector, this can be a classroom demonstration. Our students found that the diffusion dynamics provide a valuable illustration of the fractal branching process. In addition, because the materials are inexpensive and readily available, this works well as a lab exercise and as the basis for a term project.

A minor modification of the watercolor experiments yields patterns closely resembling (in form and process) the viscous fingering of a Hele-Shaw cell (chapter 4 of Feder (1988)). Here the materials needed are finger paint paper, finger paints, and blank overhead transparencies. First, apply finger paints (of rather high viscosity) near the middle of the paper. Cover the paper with the overhead transparency. Flatten and spread the paint by applying pressure to the transparency. When the desired mix of paint has been pressed thin between the paper and the transparency, pull the transparency off the paper. As the transparency separates from the paper, air invades the paint along fronts whose instability is driven by the difference of viscosities (see the discussion in chapter 4 of Feder (1988)). Branches off branches off branches form, resulting in a dendritic fractal pattern. See the right side of Figure 4. An interesting variant is to apply the paint to the palm of one's hand, press the hand to the paper, and pull the hand vertically from the paper.

This method gives a model for denseness of periodic orbits, but in a spatial pattern. Because they take the longest spatial extent to repeat, the largest repeating features correspond to the lowest period cycles. The second largest repeating features correspond to the next to lowest period cycles, and so on. By looking at this pattern, the abstract fact that higher-period cycles fill the dynamics ever more fully is relatively easily seen through the corresponding feature of the physical example. See Figure 5. The longest branches have the lowest spatial frequency (they repeat less often as we scan across the picture); the shortest branches repeat most often, so have the highest spatial frequency.

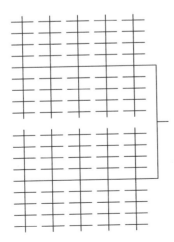

Figure 5: A spatial realization of denseness of periodic orbits.

5 Precedents

In fact, some variations of this last method have been used by several artists, including Max Ernst, specifically with the technique "decalcomania" (Janson (1991), page 731). Transferring oil paint to canvas by applying it to another surface and pressing that surface to the canvas, Ernst obtained richly-textured images evoking a dream quality. Some good examples include "Mythological figure - woman" (1940), "Three well-tempered cypresses" (1949), and "Blue mountain and yellow sky" (1959) (Quinn (1977), pages 217, 267, and 315).

Oscar Dominguez used decalcomania as early as 1935 (Passeron (1978), page 44). Placing paint or ink on a surface, he covered the surface with paper, pressed it into the ink or paint, then pulled the paper off. "Decalcomania" (1936) (Passeron (1978), page 46) is a good example.

Forms of this sort are relatively easy to produce. See Figure 6. On a canvas board place a liberal amount of gesso and some acrylic ink for color contrast. Cover with another canvas board and remove the second board. The infusion of air into the gesso and ink mix produces a pattern similar in appearance to Ernst's decalcomania.

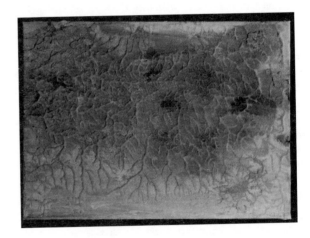

Figure 6: Decalcomania of gesso and ink.

Janson (1991), pgs 731 and 625, views decalcomania as related to a method Alexander Cozens developed from da Vinci's suggestion that artists look for recognizable shapes in stains on old walls. Cozens smoothed a crumpled sheet of paper, dropped ink blots onto it, and in the resulting patterns sought inspiration for landscapes.

The long history of using natural processes to generate subtextual patterns lends a degree of legitimacy to our inversion—using the artistic aspect of these processes to explore the properties of natural fractals. Most of our students have not been familiar with this aspect of the art of Ernst or Dominguez; pointing out this similarity has increased the effectiveness of the presentation.

6 Projects

The main motivation for the projects is to give students the opportunity to take some of the ideas of the course and bring them into their worlds. Many of our students have backgrounds in the humanities and arts, so projects often fall into these fields.

Student projects have included musical compositions, creative writing of short stories and poetry, dance, videos, and visual art. The intent, well-realized in some projects, not so well in others, is to produce an example of fractals or of chaos in the chosen medium.

Some musical compositions have been sophisticated polyphonic pieces, reminiscent of Philip Glass, in which the same basic pattern is repeated over several time scales. Others have been generated more mechanically, for example, dividing the output of the logistic map into several bins and as the time series of the logistic map unfolds, assigning notes and durations by the corresponding sequence of bins. Several students used this method with real-world data, including daily milk output from a dairy farm. This was Fractal Moosic, the Pastural Symphony.

Most creative writing projects have been efforts to illustrate chaos in daily life, primarily through sensitive dependence on initial conditions. A particular series of events is described, then a small initial change imposed and the resulting changes explored. While some of these have been quite clever, the main problem lies in understanding that chaos is indicated when small changes grow into large effects; no one is surprised when big changes have big effects.

Some of the poetry projects have involved sensitive dependence on initial conditions, others have been rather rigid attempts to build a Cantor set into the word repetition pattern, mimicking the analysis of Wallace Stevens in Pollard-Gott (1989). However, one of the most successful poems was a palindromic poem, with the palindromic pattern repeated on several levels.

The best fractal dance project we have seen had two tall students executing a slow, stately walk along a large circular path. Around each tall student, two middle height students walked briskly in circles. Around each middle student, two small students ran in circles. This was explained as fractal motion.

Videotapes included meanders around campus, pointing out natural and architectural fractals, and videofeedback experiments (Chapter 1 of Peak & Frame (1994)).

The widest range of projects has been in visual arts. Among the best have been nonrepresentational prints Patnode (1995), copper etchings, and oil paintings where different pigments have flowed into one another. While we shall not show one of the bad projects, Figure 7 is typical. Take a familiar construction from class—for example, generating the Sierpinski gasket—and apply it to something loosely fitting into the category of visual arts. This is a weak project

Figure 7: Archetype of an inappropriate project involving fractals and art. In this example, the younger brother of one of the authors undergoes iterative processing toward a Sierpinski gasket.

because it adds nothing new to the student's understanding of any of the concepts from the course. Little was learned here, except that most photocopiers don't allow you to reduce by $\frac{1}{2}$ (Figure 7 was produced by scanning a photograph, then shrinking and copying it in a paint program.) Equally unimaginative projects have included coloring a picture of the Sierpinski gasket, and coloring grid squares in a random order. Evidently, those students doing projects of the latter type did not believe our often-repeated mantra, "Chaos is not random."

Contrast Figure 7 with Figs. 8–10, good examples because they incorporate artistic processes with natural fractal dynamics.

Figure 8 shows monoprints. These images were created by painting onto plexiglass with waterbased crayons (e. g., Caran D'Ache). The crayons are first allowed to soak in water and then used to draw directly onto the plexiglass. The water acts as a vehicle for mixing and bleeding of the different pigments. After the plexiglass has dried completely, a damp piece of paper is placed overtop of the plexiglass. Running the plexiglass and paper through a press forces the pigment into the paper. Because the process of making the print removes most of the pigment from the plexiglass, only one print is produced. The most obvious fractal nature of these pieces is a result of the bleeding and mixing, natural fractal processes.

Figure 9 shows collagraphs. A collagraph is created by glueing various collage materials (common examples include aluminum foil, paper, string, and sand) onto a piece of cardboard or wood, then sealing this plate with polyurethane. The

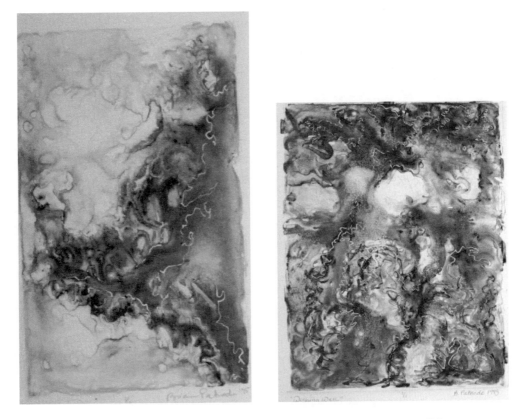

Figure 8: Fractal monoprints. Left, "Untitled"; right, "Wishing well."

Figure 9: Fractal collagraphs, both "Untitled." See the color plates.

Figure 10: Fractal papermaking, "Untitled." See the color plates.

dry plate then can be inked and run through the press. Unlike with monoprints, with collagraphs several prints can be made and different color combinations explored. Also, here the fractal nature of the printmaking process comes more from the composition than the medium. In the right side of Figure 9, notice the textures, particularly in the lower left corner, and also the boundaries between the different colored regions. These textures were produced by torching wet Elmer's glue, yielding intricate patterns of bubbles and cracks. In the left side of Figure 9, the background is tiler's cement and the figures were created by loosening the weave of pieces of burlap. In general, experimentation with different material gives students the opportunity to discover both fractal and artistic aspects of many common substances.

Papermaking is the process illustrated in Figure 10. In this example, again, it is the process that produces the most apparent fractal features. Note in particular that the edges resemble the fractal boundaries produced by tearing paper. Experimentation is limited only by the students' imaginations.

When these slides were presented in *Fractal Geometry for Non-Science Students* at Yale in 1995, the fractal aspect of the left side of Figure 9 was questioned. This gave the opportunity to explain that on first glance the piece might appear Euclidean. Its fractal nature is revealed by looking more closely within the shapes, much as is the case with nature. Another question addressed the fact, presented in the discussion accompanying the slides, that the artist produced these images before becoming familiar with the language of fractal geom-

etry. The student asked if now being comfortable with fractal geometry affected the artistic approach. In fact, the processes have not changed; the only change has been in developing a new vocabulary to describe the work. We take this as yet another example of the appropriateness of fractals in the world of art.

7   Conclusion

The last few years have seen serious efforts to incorporate writing in the mathematics curriculum. Doubtless, part of the motivation is to improve the writing skills of mathematics students. Another reason is the power of verbalization itself. By writing an approach to a problem, students have to be less lazy with their thinking. Assumptions must be spelled out, and sloppy deductions are laid bare for all to see. There is no doubt that writing strengthens the deductive aspect of problem solving.

But many problems have a visual component. Practice with visual pattern construction and recognition is an important step in the inductive side of problem solving. Thinking in pictures allows patterns to self-organize. Especially for seeing how geometry models nature, visual thinking is of paramount importance. All too often, the pictures we and our computers draw do not appear natural. Students see a gap between what we say ("Fractals are the language for describing nature.") and the sort of fractals we show them, at least initially. Incorporating simple artistic exercises allows students to produce natural fractals and so understand first-hand the differences between natural and mathematical fractals. In addition, these exercises give a clear presentation of a relation between fractals and art, putting fractals more firmly in the worlds of the humanities and arts students.

Fractals continue to provide a powerful lesson that the various fields of learning are not as disjoint as our narrow graduate training suggests. Indeed, Mandelbrot's 70th Birthday Symposium was attended by not just the usual crowd of physicists, mathematicians, computer scientists, biologists, chemists, geologists, and engineers, but also by artists, economists, and physicians. By incorporating artistic experiments in the beginning of introductory fractals courses, we can more forcefully bring to our students this message: there are profound interconnections between all branches of knowledge.

Chapter 15

Order and Chaos: Art and Magic
A First College Course in Quantitative
Reasoning Based on Fractals and Chaos

David Peak and Michael Frame

1 Why a first course in quantitative reasoning grounded in fractals and chaos?

The usual first courses in college science and mathematics model nature as if linearity were supreme and nonlinearity an ignorable annoyance. First courses are almost exclusively reductionist and employ an arsenal of analytical techniques drawn from classical algebra and calculus. These courses frequently deal with situations that run counter to common experience—for example, in the often invoked charge to ignore friction—emphasizing, instead, contrived conditions that are sterile and largely irrelevant. These courses typically proceed via a series of formal exercises that are intended, on the one hand, to strengthen the promising student's manipulative skills, and, on the other, to weed out the unpromising. Even for the best students this opportunity to develop tools and algorithmic strategies for some promised or implied later use all too often goes unappreciated. Rather than providing welcoming entry points to the study of major intellectual disciplines, these courses are perceived as barriers that have to be hurdled before getting to the good stuff. Regrettably, many beginning students never make it that far. They turn to other pursuits, forever convinced of their hatred of physics, or mathematics, or... (insert your own favorite technical discipline). Inevitably, these haters of mathematics and science become taxpayers, and sometimes they actually vote.

In contrast to linear idealizations, fractals and chaos are both descriptive and quantitative tools with firm roots in phenomena. Intrinsically nonlinear, they seem to provide much more apt and relevant models for the actual, unsanitized world. To the many students whose quantitative development level is still solidly concrete, fractals and chaos are more appealing metaphors for reality.

Use of fractals and chaos in courses on quantitative reasoning requires a shift in mind-set. First, the inherent nonlinearity of fractals and chaos frees both instructor and student to be less reliant on classical analysis. Instead of thinking predominantly in symbols and equations, fractals and chaos encourage us to think more in pictures. Obviously, use of the computer for visualization and the development of intuition is essential in this process.

Irrespective of the merit of the latter point, it is certainly true that many scholars outside of science and mathematics have applied aspects of these ideas to the understanding and development of their own disciplines (see for example, Shearer (1992), Briggs & Peat (1989), and Wheatley (1994)). Thus, use of fractals and chaos in elementary science and mathematics courses provides unusual opportunities to construct meaningful bridges between almost all the fields of study of the university, and simultaneously establishes the utility of quantitative reasoning in areas of more direct interest to many students.

In short, our case for a first quantitative reasoning course concentrating on fractals and chaos is that such a course (a) frees instructor and student from the pressure of meeting a fixed curriculum, (b) is about exciting, new ideas that span many disciplines, (c) can be filled with visually appealing material that is less abstract than that covered in traditional courses based on classical analytic models and methods, and (d) is fun for professor to teach and student to take. If, after all that, your students ask for additional rationale for why they should enroll in such a course, we recommend offering this speculative (but, likely, soon to be true, and aready supported by reports from some of our former students) answer: as applications of fractals and chaos infuse the literature and lore of many different disciplines, those students—especially in the liberal arts—whose undergraduate transcripts display a course in such matters will have an advantage in competing for employment.

## 2	Order and Chaos: Art and Magic

Motivated by many of the considerations outlined above, we began offering a quantitative reasoning course for beginners, thematically based on the concepts of fractal geometry and chaotic dynamics, at Union College, in 1988. *Order and Chaos: Art and Magic* (OCAM, for short) was designed primarily for students with no prior college-level experience in mathematics or science and with no stated interest in pursuing such disciplines for their majors. OCAM had an associated laboratory which met occasionally, as the material under discussion warranted.

From the beginning, we were conscious of the need to motivate the course discussions with practical examples. While some students find the neatness of puzzles and tricks—like playing the Chaos Game, for example—sufficient justification for their study, most of those for whom this course was developed want to know, "Why am I studying this stuff— who cares?" In response, we scoured the literature for applications of the ideas of fractals and chaos in disciplines outside of mathematics and incorporated as many of these into our presentations as we could make intelligible to a general audience. (For an update, see *Panorama of fractals and their uses* in this volume.) Some applications have economic implications. Others are of a more purely intellectual interest. Examples of applications that elicited positive responses in students include: use of fractals for image compression (Barnsley (1993), Barnsley & Hurd (1993)); consequences of fractal distributions of ores and petroleum (Carlson (1991)); consequences of fractal distributions of vegetation types over natural landscapes (Milne (1991)); fractal characteristics of music (Voss & Clarke (1975), Gardner (1992), Hsü & Hsü (1990), Hsü & Hsü (1991)) and poetry (Pollard-Gott (1986)); search for evidence of chaos in stock prices (Berreby (1993), Gilmore (1993)) and physiology (Goldberger, Rigney & West (1990)); controlling chaos in lasers (Roy, Murphey, Maier, Gills, & Hunt (1992)), electronic circuits (Hunt (1991)), mechanical devices (Ditto, Rauseo, & Spano (1990)), rabbit hearts (Garfinkel, Spano, Ditto & Weiss (1992)), rat brains (Schiff, Jerger, Duong, Chang, Spano & Ditto (1994)), and communications devices (Hayes, Grebogi, Ott & Mark (1994), Cuomo, Oppenheim & Strogatz (1993)); using chaos to create music (Stewart (1989), pages 157–161) and art (Peak & Frame (1994), pages 228–230); chaotic characteristics of textual deconstruction (pages 1–33 of Hayles (1991),

Argyros (1991)); chaos as a metaphor for history (Shermer (1993)); and complex adaptive behavior in artificial networks that learn (Allman (1990)) and in industrial organizations (Wheatley (1994)).

Variants of OCAM have also been tried by us at Yale University, the University of Richmond, and Utah State University. As a consequence, our approach to quantitative reasoning has been field tested by roughly 2000 students with widely different backgrounds, in quite different settings. We have not attempted to do a statistically careful study of the efficacy of our courses in changing student opinion about studying quantitative material. On the other hand, formal, institutional evaluation tools have been used at each of these venues. In broad swath, the general enthusiasm of the students toward their experience in our courses has been unusually positive, well above the respective departmental norms for such courses. (Selected comments from these evaluations can be found in Section 5.) More anecdotally, we know of a number of our students who subsequently took calculus specifically so that they could take more advanced courses in fractals and chaos; at least one of our students changed his major (from political science to physics) because he got so involved in his OCAM term project; and another decided to go on to graduate school to continue studying nonlinear dynamics because she so enjoyed OCAM.

The performances of our students on their exams and quizzes, the character of their reports on laboratory assignments, the quality and ingenuity of their term projects, and their comments and term-end course evaluations, have all helped refine the material we have developed for these courses. This material includes Peak & Frame (1994), software, extensive problem sets with selected answers, laboratory instructions, and example syllabi. These are available on the web at

`http://www.maa.org/Fractals/OCAM/welcome.html`

Originally, we tried using Gleick (1987) and Briggs & Peat (1989) as texts for OCAM. In short order, we found that these purely qualitative treatments were insufficient to give our students more than a fuzzy sense of the concepts. Students needed at least some quantitative detail and practice to sharpen their mastery of the material. Thus we came to write our own book, intending it to reside somewhere between the latter works and the many more mathematical texts that are available.

Reader response to Peak & Frame (1994) suggests that, if nothing else, it is at least an unusual mathematics/science book. With the needs of our (mostly non-quantitatively inclined) students in mind, we wrote Peak & Frame (1994) in an intentionally conversational style. The book aspires to reach the general through the specific. Each topic is introduced with a concrete, numerical example and illustrated with lots of photos, drawings, and graphs. Despite its breeziness, Peak

& Frame (1994) treats some genuine mathematics: the logarithm and exponential functions, graphical iteration, continuity and smoothness, the derivative, complex arithmetic, and Newton's method for finding roots. In addition, we include elements of statistical reasoning and a qualitative discussion of Fourier analysis. Sandwiched between introductory and concluding chapters filled with attention-grabbing examples and speculations, are two chapters dealing with complexity in space (covering iterated function systems and fractal dimensions), three chapters dealing with complexity in time (covering the nature of chaos and its defining properties, universality in the period doubling route to chaos, and detecting and controlling chaos), and three chapters dealing with complexity in space and time (covering the Mandelbrot Set, fractal basin boundaries, and cellular automata).

Peak & Frame (1994) includes a small subset of all of the student exercises we have developed as we have taught these courses. A much larger collection along with many answers, software that can be used in an exploratory mode to investigate iterated function systems, one-dimensional dynamics, detecting chaos, the Mandelbrot Set, cellular automata, and illustrative laboratories are available for free via the Internet. Instructions for accessing these files can be found in Appendix A. In Appendix B we list four syllabi for introductory quantitative reasoning courses based on fractals, chaos, and complexity that we have actually taught at one institution or another.

Because of the visual appeal of the material and its often counter-intuitive character, the medium and the message are more inextricably intertwined in OCAM than in most courses. We have offered OCAM in sections with enrollments ranging from 15 to 180, and in such physical settings as large lecture halls, classrooms equipped with student computer stations, and more traditional, electronically-challenged environments. In all cases, we have found it important to liberally punctuate our board presentations with multimedia events—computer simulations, audio and video fragments, and real world, live demonstrations.

At Union College, the delivery of such multimedia material is aided by a computer-outfitted classroom and a mobile multimedia station we designed and assembled (with external support from the National Science Foundation, Sloan Foundation, Pew Charitable Trust, and Apple Corporation supplementing internal funds) for this purpose. The media station includes a computer, video disk player, CD-ROM reader, VCR, and data acquisition board. The classroom can be used in a lecture-investigation mode with chairs arranged for lecture and computer stations around the periphery available for students to turn to when appropriate. As implemented at Union, the classroom is an example of a modest size *fractal classroom*, a place where learning can occur on multiple length scales: for 30 students in a lecture format, for collaborative learning teams of size 3-5, and for students working alone. While the facilities at Union are reasonably hospitable

to a fully multimedia-enhanced course, we have certainly persevered in much less amenable situations. Minimally, we have made do with a VCR and a laptop computer connected to an LCD projection device.

3 The value of a laboratory

Our experience convinces us that exploration-based laboratories are especially valuable for OCAM's intended audience, students with limited exposure to science. Not only do such laboratories bring to life the major concepts presented in the course, they also provide an arena in which the students can experience science as it is actually done. To this end, we have tried to construct exercises that encourage the students to act as scientists, making guesses, making relevant observations, and trying to explain why their guesses were right or wrong. Done well, exploration-based laboratories can reveal science as a human activity, provide an introduction to science's open-ended nature, and enhance students' sense of ownership of the material. Grappling with real data gives students an appreciation for what can be claimed as a result of an experiment, and what is beyond the range of legitimate interpretation. They begin to understand the importance of asking good questions, and the necessity of giving up a favorite conjecture in the light of contradictory experimental evidence.

Here, we give fairly detailed descriptions of two laboratories, one computer-based and one physical, that we have used with some success. We have tried other laboratories including extracting IFS parameters from a scanned leaf image, measuring properties of a simple pendulum, probing attractors and chaos in one-dimensional dynamics, observing period doubling in diode-inductor circuits and in dripping faucets, employing visual assays of periodic and chaotic behavior in various data, investigating relations between the Mandelbrot set and Julia sets, exploring the effects of changing the rules and starting conditions for one- and two-dimensional cellular automata, and training neural networks to perform simple tasks.

3.1 Iterated Function Systems laboratory

Here students use Iterated Function Systems (IFS) to explore how fractals are constructed by iterating geometrical transformations (Barnsley (1993), Hutchinson (1981)). Instructions for the lab include this statement, "You will not be graded on the correctness of your guesses, but rather on the clarity of your explanations of why your guesses were right or wrong." In addition, students are told independent explorations, beyond those described in the lab handout, are required in order to achieve an excellent grade.

Following fairly standard conventions, we represent each affine transformation as

$$x \rightarrow x \cdot R \cdot cos(\theta) - y \cdot S \cdot sin(\varphi) + E,$$
$$y \rightarrow x \cdot R \cdot sin(\theta) + y \cdot S \cdot cos(\varphi) + F.$$

A set of affine transformations can be presented as a table in which are listed the R, S, θ, φ, E, and F values of each. For example, an equilateral Sierpinski gasket is generated by the IFS

Rule	R	S	θ	φ	E	F
1	0.5	0.5	0	0	0	0
2	0.5	0.5	0	0	0.25	0.433
3	0.5	0.5	0	0	0.5	0

We use two versions of IFS, the deterministic IFS algorithm (DET) and the random IFS algorithm (RAND). Denoting the IFS transformations by $T_1, T_2, \ldots, T_N$, DET applies each T_i to a set of points and takes the union of the resulting sets. Iterating this process generates a sequence of point sets converging to the attractor of the IFS. In contrast, RAND applies the T_i in random order, one at a time to a succession of individual points. Starting with an arbitrary first point, each new point is one of the T_is operating on the previous point. The accumulated set of points produced in this way converges to the same fractal obtained by DET.

Here is an outline of the steps of the lab. The lab uses the program *TreenessEmerging*. Students are given supplementary instructions on how to handle various aspects of this program.

1. To illustrate that DET converges to the gasket regardless of the initial picture, use the mouse to draw a starting picture and observe that iteration of the IFS rules produces a sequence of pictures becoming ever more like the gasket. Running RAND with the same parameters illustrates another method of producing the gasket from these rules.

2. For the first experiment, select the equilateral gasket parameters and predict what happens when all the R and S values are changed from 0.5 to 0.333. (Make a sketch of what picture you expect will emerge.) Run RAND, compare the results with your predictions, and try to explain the differences. ("Predict, run, compare, explain" for short.)

3. Next, guess what changes of the E and F parameters will produce a right isosceles gasket. Enter these values into the parameter table, run RAND, compare the results with their predictions, and try to explain the differences.

4. Repeat part 2 with all the R and S values set to 0.667.

5. Return to the equilateral gasket parameters. Now change all the R values to 0.333, leaving the S values at 0.5. Predict, run RAND, compare, explain. Reset the R values to 0.5 and change all the S values to 0.333. Predict, run RAND, compare, explain.

6. Return to the equilateral gasket parameters. Set the second row E and F values to 0 and 0.5. Predict, run RAND, compare, explain.

7. Return to the equilateral gasket parameters. Guess what changes would produce a Cantor Middle Thirds set? Run RAND with these parameters. Compare with the desired picture. Explain any differences.

8. Return to the equilateral gasket parameters. In the first row, set θ and ϕ to 120 and change E to 0.5. Predict, run RAND, compare, explain. Why does this give exactly the same gasket? Observe that while one set of rules determines exactly one fractal, distinct sets of rules can determine the same fractal.

9. Return to the equilateral gasket parameters. In the first row, set R to -0.5. Predict, run RAND, compare, explain.

10. Run RAND with the Koch curve parameters

Rule	R	S	θ	φ	E	F
1	0.333	0.333	0	0	0	0
2	0.333	0.333	60	60	0.333	0
3	0,333	0.333	-60	-60	0.5	0.289
4	0.333	0.333	0	0	0.667	0

(and all probabilities equal to 0.25)

Now change the two -60s to $+60$s. Predict, run RAND, compare, explain.

11. Return to the Koch curve parameters. Replace the 60s in Rule 2 with 30s. Predict, run RAND, compare, explain. Return to the Koch curve parameters. Replace the -60s in Rule 3 with -30s. Predict, run RAND, compare, explain.

12. Return to the Koch curve parameters. Replace the 60s in Rule 2 with 30s and the -60s in Rule 3 with -30s. Predict, run RAND, compare, explain.

13. Return to the Koch curve parameters. Replace both the 60s in Rule 2 and the -60s in Rule 3 with $+30$s. Predict, run RAND, compare, explain.

14. Return to the Koch curve parameters. Change all the R and S values to -0.333. Predict, run RAND, compare, explain.

15. Return to the equilateral gasket parameters and run RAND. Note how the gasket fills in. Now change the probabilities to 0.6, 0.2, and 0.2. Run RAND with these probabilities, and watch how the gasket fills in. Explain.

16. Run RAND with the Tree parameters

Rule	R	S	θ	φ	E	F
1	0.05	0.6	0	0	0	0
2	0.05	-0.5	0	0	0	1
3	0.6	0.5	40	40	0	0.6
4	0.5	0.45	20	20	0	1.1
5	0.5	0.55	-30	-30	0	1
6	0.55	0.4	-40	-40	0	0.7

(See Figure 1.) Change the first address box to 1. Running the program highlights the region with address 1. Repeat, making the first address digit 2, 3, 4, 5, and 6 in turn. Describe the corresponding parts of the tree. Leave the first address digit 6 and change the second digit to 5. Describe what you see. Try a few more combinations and describe how the addresses are related to one another.

Figure 1: Fractal Tree, generated by the six function IFS of part 14 of the IFS Lab.

17. With what you have learned about IFS, develop a set of rules that yields a natural looking picture (a leaf, a bush, a river network, ...).

3.2 Dimension laboratory

In this laboratory students measure the dimensions of two physical fractals, paper wads and tear paths.

3.2.1 Paper wads

When sheets of paper are wadded up into balls, the wadding process has the potential to create a fractal structure. That structure can be expected to produce a fractal dimension somewhere between 2 and 3.

Make a series of five tightly wadded balls of tissue paper (very un-stiff), five of typing paper (medium stiff), and five of brown paper bag paper (very stiff). Before making any measurements, guess which set has the higher dimension. What reasons can you give to support your answer? In each case, collect 22 same size pieces of paper. From these, assemble five different sized pieces of paper to form the paper wads.

Tape 16 together in a 4 x 4 arrangement.

Tape 4 together in a 2 x 2 arrangement.

Take a single piece of paper.

Take the remaining piece of paper and cut it in half vertically and horizontally. Take one of the four pieces.

Take another of the pieces in the previous step, cut it in half vertically and horizontally, and take one of these four pieces.

Crumple each to form a paper wad. Try to crumple all five in the same way. After each has been allowed to settle for a minute, measure its diameter with a caliper in three different positions (try to get the longest diameter, the shortest diameter, and one in between) and compute the average diameter.

From smallest to largest, these paper wads have masses $M = 1$, $M = 4$, $M = 16$, $M = 64$, and $M = 256$. To get the corresponding values of L, divide the diameter of each wad by that of the smallest (i.e., $L = 1$ for the smallest wad). Enter these M and L values into *Cricket Graph III* or some similar graphing software with curve-fitting capabilities. Find the dimension d by fitting a power law $M = L^d$ to the data. (Alternatively, plot $\log(M)$ as a function of $\log(L)$ and find d from the slope of the straight line that best fits the data.)

Were your guesses about dimension right or wrong? Try to relate your observed results to the physical properties of the different papers you used.

3.2.2 Tear paths

When a piece of paper is torn, the tear path is rough and irregular. It writhes around in a way which is similar to a river or coastline. How close is that similarity?

Make a notch in the middle of one side of a piece of typing paper. Hold the sides of the paper firmly and pull them apart (in the plane of the paper) until the paper tears. The tear edge will be jagged and rough. Overlay the torn edge on a piece of graph paper and estimate the dimension of the edge by box counting (as described in Peak & Frame (1994)). Again, we seek a relation $M = L^d$.

Take a single square of graph paper as the unit: $M = 1$, $L = 1$.

Count the number of squares of graph paper the tear path intersects in a horizontal run of 10 squares. The number of squares you count is M and $L = 10$.

Count the number of squares of graph paper the tear path intersects in a horizontal run of 50 squares. The number of squares you count is M and $L = 50$.

Count the number of squares of graph paper the tear path intersects in a horizontal run of 100 squares. The number of squares you count is M and $L = 100$.

Enter these values of M and L into *Cricket Graph III* and find d by fitting a power law to the data. (Or use the logarithm method outlined above.)

Does the dimension depend on the speed with which the paper is torn? Repeat the previous steps for a couple of different tear speeds.

Try tissue paper and heavy packing paper instead of typing paper. Does the dimension of the path depend on the structure of the paper?

Compare your results with typical river and coastline dimensions.

4   The value of a term project

In mathematics and science courses, students have relatively few opportunities to exercise unfettered creativity. Such courses are frequently ruled by a tyranny of curriculum, a glut of concepts and techniques that are mandated by actual or supposed expectations of other courses, accreditation boards, or unnamed higher authorities. The instructor of a course like OCAM is liberated from such concerns because the only expectation his or her course must heed is to provide its students with a positive educational experience. And, because that course deals with new, rapidly evolving, and highly interdisciplinary material, it can be a perfect incubator for student imagination.

In recognition of this unusual pedagogical opportunity, our courses always require a term project. Moreover, because students will typically have to cooperate with others to solve problems in their professional lives, we strongly encourage them to polish their collaborative skills by working on their projects in small teams. In broad outline, the project is supposed to reflect the use of fractals or chaos or self-organization in some subject of interest to the team. Inasmuch as our students are predominantly liberal arts majors, that subject is often in the humanities or the arts. Each team is expected to deliver a product, a video, photographs, a painting, music, dance, or, more prosaically, an analysis of some kind. In addition, teams must submit a paper describing the underlying premises, how the project incorporates the ideas of the course, the steps taken in designing and constructing the project, and a critique of how well the team feels the project achieved the stated objectives. High grades are awarded for novelty, ingenuity, and clear evidence of mastery of the associated concepts. As measured by oral comments and in the written course evaluations, this aspect of the course has been consistently successful in generating enthusiasm and promoting engagement.

This is not to say every project is wonderful, nor have students always succeeded in finding projects based on their main interests. For example, we did not encourage students to pursue a project looking for fractal patterns in the abductions of humans by extraterrestrials. Sometimes, even soundly based projects are subject to mistakes in execution. In class we use several animations to explore the effects of parameter changes on IFS, though for clarity we use variations on the Sierpinski gasket. Several years ago a student wanted to do an animation of a fractal tree swaying in the breeze. Starting with the IFS parameters for a reasonably natural-looking tree (see Figure 1), he guessed at how to modify the angles of the branches to mimic the effect of wind blowing through the tree. Unfortunately, he did not check the intermediate stages of his animation, but simply rendered each picture and pasted it into the animation engine. The result was an animation of a tree doing jumping jack exercises (clapping its hands above its head), not a complete disaster, but not what he had intended.

Over the years we have gotten some very interesting projects. These have included:

4.1   Videofeedback experiments

Videofeedback is a form of iterative image processing. The output from a camcorder is fed through a VCR (to record the images) and into a monitor. As long as the zoom is adjusted so the image of the monitor is smaller than the monitor (so the process being iterated is a contraction), a great variety of shapes pulsates across the screen, depending delicately on the positions of the camcorder, the room light level, and the settings (brightness, contrast, ...) of the camcorder (pages 19–22 of Peitgen, Jurgens & Saupe (1992a)). For some settings, complicated, organic-looking patterns twist and writhe across the screen. Student projects have ranged from careful experiments changing one parameter at a time and recording the results, to stream-of-consciousness efforts to make the images reflect music playing in the background. Experiments of this kind are open-ended, and, approached properly, can give students a sense of the creativity and intuition necessary in scientific research.

4.2   IFS pictures

After being exposed to the powerful way IFS can generate complicated-looking images from small sets of rules, some students try to use IFS to create art. This gives them direct experience with how changing the parameters affects the picture, much more so than watching us perform similar experiments in class. Projects have included photographing trees and finding the IFS parameters to mimic the photographs, finding IFS rules to produce each letter of the alphabet (obviously not for text compression, but for artistic effect, an H made up of smaller Hs, each made up of still smaller Hs, ...), and a recognizable caricature of a student's face.

4.3   Fractal and chaotic poetry

Students have been particularly interested by the analysis of repetition patterns in the poetry of Wallace Stevens presented in Pollard-Gott (1986). The method is simple. Assemble a linear array of cells, one for each word of the poem. Associate the left-most cell with the first word of the poem, the next cell with the second word, and so on. Now select a key word, a repeated word having some significance to the reading, and darken each cell corresponding to this word. Surprisingly, the pattern that appears often resembles the first few stages in the formation of a Cantor set. Students have mimicked this analysis with other poets, using repetition of sounds or ideas instead of key words, and with poetry in languages other than English. As one would expect, the results are mixed. Some poems appear to have a fractal repetition structure when viewed in the right way, some don't. Nevertheless, the existence of fractal structures, obviously not consciously designed by the poets, makes an impression on students. In related projects, students have tried to write their own poems with a fractal structure built in. These, too, have been successful in varying degrees. One of the best was a fractal palindromic poem. The poem exhibited palindromic structures on three levels. Another interesting poetry experiment involving chaos resulted from decorating the sides of the tent map with evocative words and generating a kind of artificial haiku by iterating from some arbitrary starting word.

4.4   Fractal and chaotic music

Fractal aspects of music are familiar in the literature (Voss & Clarke (1975), Hsü and Hsü (1990, 1991)). Musically inclined students have produced a variety of fractal compositions, some more rigidly exhibiting the same structure over several levels, some less so. The best have sounded like the early work of Philip Glass. Producing these compositions has helped students grasp the multiple levels of fractal structures in a different way. With fractals in space we can view all the levels simultaneously, while with fractals in time (and the fractality of music inheres in its time evolution) we perceive the multiple levels only in our minds, having heard the notes in sequence.

Chaotic music can be extracted by assigning notes and durations to natural or synthetic time series. One memorable example used the daily milk weight data from a student's family farm. The result was entitled *Moosic: A Pasture-al, Pasteur-al Symphony*. The choice of performance has been affected by the students' particular musical talents. Keyboards and vocals are common, but we have probably heard the only fractal composition for solo trombone.

4.5   Fractal and chaotic dance

Several groups of students have illustrated concepts from the course by dance. One performance consisted of dancers moving in two large groups, each of which broke into two smaller groups whose members repeated the motion of the groups on a smaller scale and in faster time. In the chaotic dance, performed by two people, the dancers returned again and again to their starting positions, but with slight variations (pointed out in the accompanying text—some of the changes probably would have been too subtle to notice without this guidance). These variations unfolded in a nonlinear way, amplifying to greater and greater changes, illustrating sensitivity to initial conditions characterizing chaos.

4.6   Chaos in life

Many of our students have been fascinated by expressions in literature (e.g., Crichton (1990)) of instances of sensitivity to initial conditions: how a small change can grow to have a large effect. Indeed life, the most real text, is influenced in ways we can't predict by a seemingly insignificant or unconnected decision. Some students have taken this as a theme for creative writing projects. We have encouraged parallel developments, where the student describes the consequences of a small choice and then repeats the development, outlining what would have happened with a different choice. Some have been quite inventive in the mechanism for expressing this development, dream sequences and conversations with others being used frequently. Usually, these have been depressing stories, perhaps reflecting the unease engendered by this realization of life's chaos.

4.7   Fractal painting

Projects ranged from straightforward melding of Koch curves and Sierpinski gaskets, to abstract paint swirls and copper etching. In the former, students designed the particular details of the fractal structure of the paintings; in the latter, they followed natural processes and pointed out the fractal forms resulting from mixing and etching.

4.8   Fractal data analysis

A quick and dirty, but nonetheless interesting, way to observe correlations in a noisy data sequence is to make a *driven IFS* picture as follows (Stewart (1989), pages 222–228 of Peak & Frame (1994)). Construct from the data sequence a sequence of 1s, 2s, 3s, and 4s by partitioning the raw data into four equal size bins, where bin number 1 includes the set of data with the smallest values, bin number 2 the next smallest, and so forth. Define four transformations by $T_1(x, y) = (0.5x, 0.5y)$, $T_2(x, y) = (0.5x + 0.5, 0.5y)$, $T_3(x, y) = (0.5x, 0.5y + 0.5)$, and $T_4(x, y) = (0.5x + 0.5, 0.5y + 0.5)$. Take as a starting point $x = 0.5$, $y = 0.5$. Generate, from this starting point, a sequence of new points by successively applying the transformations that correspond to the sequence of binned data. In other words, if the first binned data value is 3, apply T_3 to $(0.5, 0.5)$, yielding $(0.75, 0.75)$; if the next binned data value is 2, apply T_2 to $(0.75, 0.75)$, yielding $(0.875, 0.375)$; and so on. Figure 2 shows an example of several years of successive differences of bond market data.

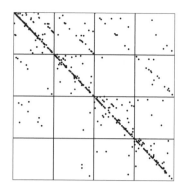

Figure 2: IFS driven by differences in daily values of the bond market.

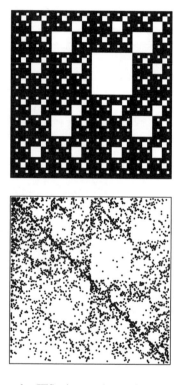

Figure 3: Top: the IFS picture determined by the condition "1 never follows 4." Bottom: the IFS driven by the DNA sequence coding amylase.

If the original data sequence is uniformly random, then this process will eventually densely fill the unit square. If, after binning, the data sequence contains dynamical correlations that result in restrictions—such as "1 can never follow 4"—the process will lead eventually to a picture inside the unit square that contains holes (regions of no fill) on many different length scales. (See the top of Figure 3.) Students have applied driven IFS to look for dynamical exclusions in financial data, meteorological data, scores of games from the National Basketball Association, sequences of camera shots (long, medium, close, and extreme close) in films, and even

patterns determined by Gematria (a family of numerological systems, dating back to the 8th century BC, for interpreting the Hebrew Bible) to test the authorship of the chapters of Genesis. While the results were not always so clear, students got a sense of the details of doing real science, examining data for removable trends, interpreting results from noisy signals, dealing with the finiteness of the data set, and the effect of modifying the bin boundaries.

This last issue, how the data are divided into bins, is especially delicate, but can be avoided if the data happen to fall into four categories naturally. DNA sequences are good examples. Associating T_1 with C, T_2 with A, T_3 with T, and T_4 with G, the bottom of Figure 3 shows the IFS driven by the sequence encoding amylase. We might speculate that the few points in the G follows C square are an indicator of the 3-dimensional structure, but this conjecture is unlikely to be tested soon. For comparison, Figure 4 shows the IFS driven by the g1564a209 and g0771a003 from chromosome 7. The sequences were obtained from the University of Washington human genome site

http://www.genome.washington.edu/UWGC/

These pictures are suggestive, but unambiguous interpretations remain to be found. Nevertheless, this way of visualizing data has attracted many student experiments.

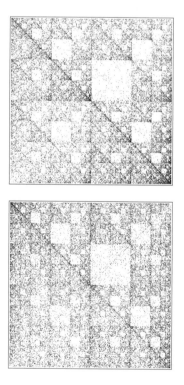

Figure 4: Top: the IFS picture driven by the DNA sequence g1564a209. Bottom: the picture for g0771a003.

5 What some students say

To obtain the students' point of view about the success of various parts of the course and the impact of the ideas presented, sometimes we added to the course evaluations this question: "Describe what parts of the course you found most interesting. Are you able to see any of these ideas in the world around you? Have any influenced the way you think about your surroundings?" Here is a sample of the replies.

"Have they influenced the way I think?? Earlier this week I was writing a paper on decorative patterns in Islamic architecture, and all I could think of was fractal geometry! Did you know that in some Islamic buildings there are big arches, and little arches with the same proportions, and then arch-shaped decorative patterns on the walls? I was reading one book that kept talking about the many 'layers of pattern' within a single mosaic. I started comparing those patterns to broccoli in my paper, but I decided it wouldn't go over too well in the Religious Studies department."

"When I enrolled in this course I thought I was going to learn that the molecules of frogs looked frog-like, and that mountains were composed of aggregates of mountain-shaped crystals and rocks. That has been both corrected and elaborated. The mystery of 'just how deeply can we go into a natural fractal before self-similarity disappears' is still the most intriguing to me. Until then, my subject to observe will be art, because: art (may) exhibit the mathematical construction that will allow for self-similarity across a wider range than for a natural fractal, yet art (should!) imitate nature, which so far shows a narrower threshold of self-similarity. Therefore, in the study of fractals, Art lies between math and nature in exhibiting self-similarity."

"I was really impressed by the subtle combinations of simple and complex behavior that seem to be everywhere. I say subtle because some situations appear to be simple and others appear to be complex. However, closer inspection reveals that many systems include both chaos and order. Think of the intricate yet orderly behavior of the magnets and the pendulum, or the windows of periodicity within the chaos of the Bifurcation Diagrams. This class has taught me to look at the world a bit more closely. Things that seem to be completely random may have some level of order, and systems that seem to act predictably may have some hidden aspects of chaos. Thanks for making this a fun and interesting class!"

"I found the in-class use of computer technology to be the most effective manner of presentation. My roommates and I have been spending the year looking around us for fractals. We've noticed them everywhere, especially outside. This has affected my view of the world somewhat—things appear more ordered and simpler than before—from the stock market to ice to patterns of flowers on the ground. I know fractals can't explain or describe everything, but they do explain a lot and provide an interesting alternative point of view of the world around us. Most interesting was the discovery of the unexpected—the bifurcation diagram, the Sierpinski gasket in cellular automata, the Mandelbrot set in Newton's method, etc. Fascinating stuff, Thanks for a great course."

"Since taking this course, every time I see something exciting I yell out 'It's fractals!' and then, usually with the help of classmates, I proceed to figure out in what way frost on the windowpane or the universe or the Communist Party is a fractal. ... I want my littlest actions to be consistent with the trends of my overall life for peace of mind. I also find the statement that there are 'windows of order' in chaos to be very hopeful. (I wonder why fractals have satisfied such emotional needs?) Thanks for everything!"

"The primary way in which this course has changed my view of the world is in the experience it has given me to the intricate (and amazing) delicacy of nature. The notion of the fragility of order I found to be most interesting. It is a concept that affects everything in my world. The knowledge of its presence in this world is frightening and exciting."

"I found the Mandelbrot set itself quite interesting. From such a simple equation, one finds so much that is hidden, seahorse and elephant valleys. It seems like a miracle to me, a great hidden agenda (hidden by whom I'm not sure—Nature itself, God). I also find the idea of sensitive dependence on initial conditions quite logical, but quite fascinating. If I ever go to philosophy grad school, I would like to write a paper considering Leibnitz as a fractal philosopher."

"The most fascinating concept about this course was chaos, its origins and its ability to create order. The world seems filled with orderly things, patterns which I've always taken for granted, not thinking about how those patterns arose from chaotic systems. I am a competitive sailor and I've always been fascinated by weather patterns and also the often complete unpredictability of systems. I've taken courses and read tons of books and articles on how to understand the weather. I've also often cursed the weatherman for being so wrong at times. I realize now why the weather just can't be predicted very accurately. Exploring fractals was amazing, too. I never even knew what a fractal was before I took this class. Now I can recognize fractals all around!"

"The wonderful thing about this class is how much it relates to my experience of the world, and how it serves as a lens for understanding it, or at least understanding its complexity. Chem 115, for example, bored me because I didn't see the application immediately. I didn't see the molecular structure of carbon and it didn't impact me on a daily level. However, with fractals, now when I look at a tree of a piece of music, a poem, a leaf, a fractal light bulb goes on in my head. I've always thought life was random in a good way, and now I feel less of a nihilistic, wandering lost soul."

"I most enjoyed the paper. It gave me a chance to apply the work done on sensitive dependence on initial conditions in class to a possible real world relationship. This class helped me philosophically. For a long time, I wondered if we lived in a solipsistic universe. Now I know that, although the only order is emergent, fragile, and unstable, we are all interconnected in such a complex pattern of events and experiences, we could not possibly hope to understand it. But we know it's there, this sensitive dependence on one another. Thank you."

"I found IFS generation and cellular automata to be the most interesting topics. Both topics deal with the reapplication of simple rules to produce (unpredictably?) complex patterns. This idea appeals to me as a way to envision the physical evolutionary process as well as from an artistic point of view. IFS generation of fractals seems to supply an artistic technique of generating self-similarity at many scales in a work of music. The cellular automata rules seem to indicate that the atomic behavior of a work of art need only (blindly) follow a simple rule, that (magically) produces wide-ranging order larger then the consciousness (read neighborhood) of the progenitor cell."

"One of my favorite parts of this course was when we studied natural fractals and examined such phenomena as the shape of snowflakes, flowers, etc. I also was impressed with the lessons of the Mandelbrot set. However, I also found the last class lecture on music rather interesting. I liked when our study of fractals interacted with the real world. ... I have found that every time I listen to music now or see growth of certain flowers, plants, etc., I find myself searching for the fractal (if, of course, there is one)."

"I was amazed at the grace by which iterative processing helped the small and simple grow into the complex and grandiose in all of the concepts introduced this term."

"The most interesting concept is chaos, as opposed to random. I have sometimes wondered if the universe was just created and left to run its coures entirely randomly, or if nature, even though it appears random, may have some order to it. After this class I am inclined to believe there is indeed some order in the world."

"What just happened [a long written discussion of an interesting exercise], the way I worked through that problem, past the apparent difficulty, is the reason I took this course. This course has greatly improved my mind."

6   Beyond OCAM

Concurrently with OCAM, we developed a junior-senior mathematics course on fractals. Though roughly the same topics are treated, the coverage is at a much more mathematically sophisticated level. For example, convergence of the IFS is established by applying the contraction mapping theorem to the space of compact subsets of the plane with the Hausdorff metric. Student projects have included self-organization in three dimensional cellular automata, fractal properties of limit sets of circle inversions, and studying the chaotic oscillations in the Belousov-Zhabotinskii reaction. Enrollments have consistently been 20 to 30 for the past eight years.

In addition to courses specifically designed to teach about fractals and chaos, we find these ideas have been making their way into other, established courses. Some examples are fractal geometry of affine transformations in linear algebra, chaos in differential equations, chaos in classical mechanics, issues of complexity in statistical mechanics, iteration of rational functions in complex analysis, the Hausdorff metric as an example of a metric in real analysis, a survey of fractals in geometry, and Newton's method and fractal basin boundaries in calculus. Typically, students have been excited by these additions, sometimes seeing them as the high point of the course. In part, we believe this follows from the more experimental aspect of the mathematics, a real departure from the usual didactic style of presenting mathematical topics. But also, fractals and chaos speak directly to the natural world, so we expect they will become thoroughly integrated into many courses. The striking visual appeal of fractals is effective in capturing students' attention. That fractals and chaos have immediate applications to the world, and can treat non-idealized models, contribute to the sense of honesty regarding applications. The youth of these topics brings home, in a very powerful way, the point that mathematics is not a finished subject, that new science can speak directly to everyone. One of the most exciting pedagogical aspects of fractals and chaos is their destroying the view that new science and mathematics requires graduate training to appreciate. This is science on a human scale—not quarks, not quasars—giving students access to applications in direct aspects of their lives.

Our hope is that OCAM eventually will depart from colleges and find permanent homes in the high school curriculum. If anything, this is more important than having OCAM in college. In high school we can reach many more students before science becomes an anathema. While fractals and chaos don't excite everyone, these topics do have a natural appeal for many students. We see the opening of a new window into students' imaginations, a new point of contact between what they learn and what they see in the world around them. We believe this is a rare opportunity to revitalize science and mathematics education, to illustrate more clearly than before the relevance of science to daily life.

Appendix A—OCAM materials available via the Internet

The course material described in this article is available on the world wide web at

`http://www.maa.org/Fractals/OCAM/welcome.html`

Here are found five pieces of downloadable executable Macintosh software, and instructions:

TreenessEmerging (150K) - iterated function systems, deterministic and random algorithms, and chaos game.

IterateAgainSam (149K) - one-dimensional dynamics, graphical iteration, histograms, time series, and bifurcation diagrams.

DesperatelySeekingChaos (152K) - data analysis by driven IFS, near returns, Kelly plots, and music.

WaitingForMandelbrot (139K) - Mandelbrot and Julia sets.

Cellabration (159K) - one- and two-dimensional cellular automata.

Syllabi for several courses based on this material, Labs exploring topics of the course,

Exercises and selected solutions, Movies (animated gifs), Sample student projects, and a small collection of links

Appendix B—Example syllabi

The following are example syllabi for several different introductory quantitative reasoning courses based on fractals, chaos, and complexity. Except where otherwise noted, the suggested readings are from Peak & Frame (1994).

I. A syllabus for a one semester-length course (that is, one with 42 sessions, each 50 minutes long) surveying the principles and applications of fractals, chaos, and complexity (with possible associated laboratories) for the beginning student is:

1. Background (3 sessions, read Ch 1): overview of ideas in course

2. Iterated Function Systems (5 sessions, read Ch 2): what affine transformations do to a picture, making pictures from nature using IFS

    (Labs: IFS explorations—guided parameter play; IFS code for a leaf)

3. Fractal dimensions (5 sessions, read Ch 3): dimension as exponent, information contained in dimension

    (Labs: dimensions of paper wads, grain clusters, tear paths, and viscous fingers)

4. Dynamics: linear and nonlinear (6 sessions, read Ch 4): what a model is, exponential growth and decay in linear dynamics, the tent map, the characteristics of chaos, the bifurcation diagram

    (Labs: properties of the simple pendulum; exponential growth)

5. Dynamics: period doubling (5 sessions, read Ch 5): the logistic map, fractal character of the bifurcation diagram, Feigenbaum's number and universality

    (Labs: dripping faucet; diode circuit)

6. Noise and detecting chaos (3 sessions, read Ch 6): statistical characterization of a noisy signal, randomness versus chaos, return maps, close pairs analysis, IFS analysis, relation of noise to art and music

    (Lab: return map, close-pairs, and IFS for some real world data)

7. Controlling chaos (3 sessions, read Ch 6): controlling the Tent Map; controlling higher dimensional chaos: magnetic ribbons, lasers, electronic circuits, chemical reactions, flames, hearts, and brains; communications; synchronization of chaotic systems: encoding communications

8. Cellular automata (3 sessions, read Ch 9): introduction to 1- and 2-dimensional CA as models for space-time structures, Wolfram's classification scheme

    (Lab: CA exploration)

9. Game of Life, self-organization (3 sessions, read Ch 9): gliders and other complex structures; self-organized criticality, Life, and colored noise

    (Lab: avalanches in grain piles)

10. Mandelbrot Set (3 sessions, read Ch 7): M-map as a transporter machine, characteristics of the M-set

11. Fractal basin boundaries (3 sessions, read Ch 8): fractal entanglement of basins of attraction

    (Lab: basins of attraction for the magnetic pendulum toy)

    OR

11. Project reports by students and wrap-up (3 sessions, read Ch 10)

Naturally, some of the sessions in this course will have to be devoted to quizzes, tests, problem solving, review, and the like. This material is ideal for student projects and we have had good success in engaging students in creative and novel explorations of the course's ideas.

II. As an alternative, here is a syllabus for a somewhat more quantitative one semester-length course treating the principles and applications of fractals and chaos:

1. Background (3 sessions, read Ch 1): same as above

2. Iterated Function Systems (6 sessions, read Ch 2): same as above + a more careful description of coordinate transformations (possibly introduce matrices)

3. Fractal dimensions (6 sessions, read Ch 3): same as above + the logarithm function

4. Dynamics: linear and nonlinear (7 sessions, read Ch 4): same as above + extension to the limit of continuous change and the exponential function (possibly introduce the derivative)

5. Dynamics: period doubling (5 sessions, read Ch 5): same as above + the derivative as the slope of the tangent line

6. Noise and detecting chaos (3 sessions, read Ch 6): same as above + harmonic functions (possibly introduce the idea of Fourier analysis)

7. Controlling chaos (3 sessions, read Ch 6): same as above

8. Mandelbrot Set (4 sessions, read Ch 7): same as 9. above + complex arithmetic

9. Fractal basin boundaries (3 sessions, read Ch 8): same as 10. above + Newton's method for root finding

10. Project reports by students and wrap-up (2 sessions, read Ch 10)

III. Here is a syllabus for a one semester-length course treating the ideas of complex adaptive systems in some depth. An additional text, such as Waldrop (1992) is recommended:

1. Overview of dynamics (4 sessions, read Ch 4, 5): attractors, chaos, period doubling, bifurcation diagram of the logistic map

2. Cellular Automata (4 sessions, read appropriate sections of Ch 9 + Waldrop (1992) Ch 6): cellular automata as models for physical, biological, and social phenomena; 1- and 2-dimensional binary cellular automata; the effects of boundaries

3. Conway's Game of Life (4 sessions, read appropriate sections of Ch 9 + Waldrop (1992) Ch 6): gliders and other complex structures, Life as a universal computer

4. Wolfram's classification scheme (3 sessions, read appropriate sections of Ch 9 + Waldrop (1992) Ch 6): examples in 1- and 2-dimensions, Langton's order parameter and similarity to logistic parameter

5. Coupled map lattices (2 sessions): independent versus collective behavior, global pattern formation from local rules

6. Self-organization (4 sessions, read approp. sec. of Ch 9 + Waldrop (1992). Ch 8): self-organized criticality (SOC); SOC in Life, sandpiles, earthquakes, and the stock market; how SOC amplifies external stimuli; $1/f$ noise and SOC and relation to esthetic response

7. Neural networks (6 sessions, read approp. sec. of Ch 9 + Waldrop (1992) Ch 3): computation by a cellular automaton, supervised learning: the backpropagation algorithm, emergent properties and the amaz-

ing leap to generalization, competitive learning: pattern recognition networks, recurrent (Hopfield) nets: associative memory

8. Complex adaptive systems (9 sessions, read approp. sec. of Ch 9 and of Waldrop (1992)): flocking (schooling) behavior in artificial birds (fish); genetic algorithms: survival of the fittest; defined fitness versus coevolution; coevolutionary forces: emergent behavior in Ray's Tierra, Holland's Echo, Arthur's artificial economy, modern business organizations

9. Organization, Complexity, and the Second Law of Thermodynamics (3 sessions, read approp. sec. of Ch 9 + Waldrop (1992) Ch 8)

10. Project reports by students (3 sessions)

IV. Finally, here is a syllabus for a much more interdisciplinary approach:

1. Overview of fractals, chaos, and complexity (3 sessions, read Chs. 1 and 10): quick introductions to self-similarity, chaos, and self-organization

2. Self-similarity and iterative processing (3 sessions, read Ch 3): examples from Nature, art, music, and poetry

   (Lab: analysis of fractal aspects of poetry, including creating some)

3. Fractal dimensions (3 sessions, read Ch 3): dimension as exponent, information contained in dimension

   (Labs: dimensions of paper wads, grain clusters, tear paths, and viscous fingers)

4. Dynamics: linear and nonlinear (6 sessions, read Ch 4): what a model is, exponential growth and decay in linear dynamics, the tent map, the characteristics of chaos, the Butterfly Effect

   (Labs: properties of the simple pendulum; exponential growth)

5. Chaos, the arts, and the humanities (3 sessions, read approp. secs. of Ch 6): the Tent Map and the Myth of Sisyphus, chaos and esthetics: a balance of surprise and regularity, chaos and art: the work of Ellsworth Kelly, chaos and music: constructing music from noises, chaos and deconstruction: postmodern physics, chaos and history, philosophical issues raised by chaos

   (Lab: creating [music, art, poetry, . . .] with chaos)

6. Controlling chaos (3 sessions, read approp. secs. of Ch 6): controlling the Tent Map; controlling higher dimensional chaos: magnetic ribbons, lasers, electronic circuits, chemical reactions, flames, hearts,

and brains; communications; synchronization of chaotic systems: encoding communications

7. Cellular Automata (9 sessions, read Ch 9): cellular automata as models for physical, biological, and social phenomena; 1- and 2-dimensional binary cellular automata; the effects of boundaries; Conway's Game of Life: gliders and other complex structures; Life as a universal computer; Wolfram's classification scheme: examples in 1- and 2-dimensions; Langton's order parameter

(Lab: CA exploration)

8. Self-organization (3 sessions, read Ch 9): self-organized criticality (SOC); SOC in Life, sandpiles, earthquakes, and the stock market; how SOC amplifies external stimuli; $1/f$ noise, SOC, and neuronal esthetics

(Lab: avalanches in grain piles)

9. Neural networks (3 sessions, read Ch 9): computation by a cellular automaton; supervised learning: the backpropagation algorithm; emergent properties and the amazing leap to generalization

(Lab: training a backprop network)

10. Complex adaptive systems (3 sessions, read Ch 9): flocking (schooling) behavior in artificial birds (fish); genetic algorithms: survival of the fittest; defined fitness versus coevolution; coevolutionary forces: emergent behavior in Ray's Tierra, Holland's Echo, Arthur's artificial economy, modern business organizations

11. Project reports by students (3 sessions)

Chapter 16

A Software Driven Undergraduate Fractals Course

Douglas C. Ravenel

The ambiguity in the title is intentional; the phrase "software driven" refers to both the curriculum of the course and the students taking it. For the past several years at the University of Rochester I have taught a sophomore level course on the mathematics of fractal images. The classroom I use was equipped with 25 50MHz 486 IBM compatible PCs, and an 8 foot overhead projector. Students taking the course are required to have a year of calculus and some computer experience. The second requirement appears to be unnecessary, because anyone wanting to take such a course would be computer literate, in many cases more so than I am. The course appears to have more appeal to computer science majors than to math majors. The level of enthusiasm in the students is rare for an undergraduate math course; on many occasions they have surprised me with their energy and originality.

1 Software used

Before describing the content of the course, I will describe the software used. All of it, with the exception of *Mathematica* and some programs I wrote, is freely available on the internet. I list them in approximate order of their importance in the course.

(1) *Fractint* is an extremely well-written omnibus program for creating fractal images quickly, and the workhorse program for the course. (Its existence accounts for the absence of commercial fractal software for MSDOS. Versions for XWindows and for the Macintosh are available now.) The program includes code for 32-bit integer arithmetic, which runs quite a bit faster than comparable floating point arithmetic and gives the program its remarkable speed. It has a comprehensive menu of fractal types, enabling one to reproduce almost every fractal image that has appeared in print anywhere. Also, it has a versatile programming language enabling a user with mod-

est skills to create many types of fractals not on the menu. Each image produced can be rotated, reflected, and magnified by many orders of magnitude, and one can manipulate the color scheme in various ways. For more information, see Peterson & Wegner (1991).

(2) *Cobweb*, *Orbit*, and *Curves* are three homemade BASIC programs which illustrate the behavior of dynamical systems related to Mandelbrot and Julia sets. In Cobweb, the diagonal line $y = x$ is plotted, along with the function $y = f(x) = x^2 + c$, for a specified real value of c. For a specified x_0 (usually 0), a vertical line is drawn from $(x_0, 0)$ to $(x_0, f(x_0))$, followed by a horizontal line to $(f(x_0), f(x_0))$. This process (vertical line to the curve followed by horizontal line to the diagonal) is repeated a specified number of times, 25 iterations being the default. The function f can be replaced by any of its iterates. With this program students can produce graphic illustrations of stable and unstable fixed points, periodic and preperiodic points, bounded and unbounded orbits, and chaos. The Curves program plots several iterates of f simultaneously. The Orbit program plots orbits for complex values of c and x and has similar features.

(3) *Mathematica* is a powerful program, especially good for symbolic and numerical computations. For example, I use it to locate periodic points near a given preperiodic point in the Mandelbrot set, which in turn are located by solving certain polynomial equations either numerically or symbolically.

(4) *FDesign* is a mouse driven program for generating IFS attractors geometrically. The user specifies a collection of contracting affine maps (an iterated function system) by drawing a reference triangle and then drawing its images under the various maps. The program displays the corresponding IFS attractor instantly, either in a corner window or on the full screen, with the images of the attractor under the various

maps coded by color. The parameters of the affine maps can be exported to Fractint.

(5) *Animation software* is used for animating a sequence of images produced by Fractint. For example, we used *DTA* to assemble a sequence of **gif** files (**gif** is the graphics format used by Fractint to store images) into an **fli** or **flc** file (**fli** and **flc** are animation formats). These can then be viewed as an animation by a suitable display program.

2   Organization of the course

I assign two texts for the course, Barnsley (1988) and Lauwerier (1991), but I do not follow them closely. (Devaney (1992) appears to be closer to the mark, but I have not had a chance to use it yet.) I use Barnsley (1988) only in the second half of the course when I talk about iterated function systems. I place on reserve most of the fractals books in our library (our librarian tells me that they are the ones most often stolen), and occasionally assign readings from them.

I begin the course by showing the video Peitgen, Jurgens, Saupe, & Zahlten (v1990). This serves as an introduction to both the subject as a whole, and to the first two major topics in the course: the Mandelbrot set and the Julia sets associated with the function $f(z) = z^2 + c$. Then I introduce the relevant concepts from dynamical systems, using the BASIC programs described above to provide illustrated examples.

The second half of the course is devoted to iterated function systems (IFSs), which is the subject of Barnsley's book. This entails discussing 2×2 matrices and metric spaces, which I do as informally as possible. I stress that the proof of what I call the main theorem of the subject (see Section 4) contains in it the ideas behind the computer algorithms used to depict IFS attractors. Such a direct link between a rigorous proof and an efficient algorithm is rare in mathematics.

The course work consists of four projects, which I encourage students to do in groups of up to three people. I offer a long list of possible topics for each project and encourage the students to invent their own topics. Each project involves some degree of programming, and I have found that they rarely need my help with that aspect of their work. Each is handed in as a printed report with computer generated illustrations, and often is accompanied by program listings.

Early on I tell the students how to use the software to make an animated sequence of fractal images; this is explained in more detail in Section 5. Everybody is required to make at least one animation as part of a project, and they seem to like doing this. The course ends with a fractal film festival of the films made by the students, with prizes offered for the best, as determined by class vote.

3   Mandelbrot and Julia sets

Here I will give a brief description of the Mandelbrot set and Julia sets. Several weeks of class time are spent on these topics.

For an analytic function such as $f(z) = z^2 + c$, the filled Julia set K_f is the set of points z with bounded orbits under iteration of f. Computationally one needs an *escape criterion* to recognize an unbounded orbit. For $f(z) = z^2 + c$, the orbit of z is unbounded if $|z| > 2$. To create an image of K_f, a program such as Fractint must compute the orbit for the value of z corresponding to each pixel on the screen. The orbit is computed up to a specified maximum number of iterations, typically 150. As soon as a value with modulus greater than 2 is reached, the pixel is colored according to the escape time, the number of iterations required to meet the escape criterion. If the modulus is still less than 2 after the maximum number of iterations, the program assumes that the pixel is in K_f and colors it accordingly. Fractint has various shortcuts for recognizing in advance when certain points have bounded orbits, thereby saving itself the trouble of computing the full 150 (or more) iterations in many cases.

A theorem of Julia and Fatou (proved before 1920, without the the aid of Fractint) says that for polynomial f, the set K_f is connected if and only if it contains the critical points of f, i.e., if and only if all the critical points have bounded orbits. For $f(z) = z^2 + c$, there is only one critical point, namely $z = 0$. The Mandelbrot set for a one (or more) parameter family of analytic functions (such as the family $z^2 + c$ for varying c) is defined to be the set of parameter values for which each critical orbit of the corresponding function is bounded. The boundedness of an orbit can be determined computationally with the help of an escape criterion as above. In a picture of the Mandelbrot set, the points on the screen correspond to parameter values, and they are colored according to the escape time of the associated critical orbit or orbits.

Let $\mathcal{M}$ denote the usual Mandelbrot set, the one for $f(z) = z^2 + c$. The first thing one sees upon looking at it is a cardioid shaped region around $c = 0$. This is called the *main cardioid*. Simple calculations show that for each c inside this cardioid, the critical orbit converges to a stable fixed point, and for $c = 0$ the orbit itself is fixed. To the left of the cardioid is a circular region of radius 1/4. The critical orbit for its center, $c = -1$, is a cycle of period 2, and for each c inside the circle, the critical orbit converges to such a cycle. Also attached to the main cardioid are infinitely many smaller regions usually called buds, and roughly circular in shape. For each value of c in a bud, the critical orbit converges to a cycle of the same period. There is one such region for each rational number between 0 and 1, the denominator of the number being the period of the cycle. There is a precise formula for the point where each bud is attached to the main cardioid, and an approximation for the size of each bud.

Closer inspection reveals these features are surrounded by a halo of many miniature replicas of $\mathcal{M}$ known as baby $\mathcal{M}$s. The largest of these has the cusp of its cardioid at $c = -7/4$, is roughly 1/50th the size of the original $\mathcal{M}$, and the critical point for the center of its cardioid belongs to a 3-cycle. Further investigation shows that these baby $\mathcal{M}$s occur in clusters around values of c for which the critical orbit is *preperiodic*, i.e., it becomes periodic after some initial transients. Used as a numerical tool, Mathematica can be of great help in locating and analyzing these preperiodic and periodic points in $\mathcal{M}$. A theorem of Lei (Lei (1990)) says that in a small neighborhood of a preperiodic value of c, the Mandelbrot set $\mathcal{M}$ exhibits self-similarity with a predictable scaling factor, and that this neighborhood is nearly identical to a similar neighborhood of the point $z = c$ in the Julia set corresponding to c. I will explain below how this result can be illustrated with Fractint.

The combinatorics of the Mandelbrot set are best encoded in Douady–Hubbard's theory of external angles, which assigns one or more real numbers between 0 and 1 to each point on the boundary of $\mathcal{M}$. The external angles of preperiodic points are always rational. Informal accounts of this theory can be found in Chapter 5 and Douady's paper in Peitgen & Richter (1986), and in Douady (1986). So far I have not had much luck in enticing students to pursue this subject in their projects.

We can consider also Julia sets for other functions, and Mandelbrot sets for other families of functions. Here are some interesting examples.

(1) Take the cubic function $f(z) = z^3 - 3az + b$, with critical points at $z = \pm a$. Fractint can be programmed to look at both critical orbits (for given values of a and b) and to end its computation when either orbit escapes. This is a 2-parameter family of functions, so the associated Mandelbrot set is 4-dimensional. A 2-dimensional slice can be viewed by imposing some relation between the two parameters—for example, holding one of them constant. Interesting animations can be made by varying this slice.

(2) Next consider the rational function $f(z) = c(z^n + z^{-n})$ for an integer n. Each $2n$th root of unity ω is a critical point, but each of the $2n$ critical orbits behaves in the same way. Hence the Mandelbrot set (with c as parameter) can be obtained by looking just at the orbit for $z = 1$. It has a garland like appearance with $2n$-fold symmetry, with a copy of the original $\mathcal{M}$ at each $c = \omega/2$, and a hierarchy of smaller copies. Each Julia set has a similar appearance.

(3) For each function $f(z)$, the associated Newton function is

$$N_f(z) = z - \frac{f(z)}{f'(z)}.$$

Newton's method for finding roots of f consists of iterating N_f: for most initial values of z, the orbit converges rapidly to a root. The *Newton-Julia set* of $f(z)$, an interesting fractal, is the set of initial values for which Newton's method fails to converge to a root. To display this fractal, we color pixels according to *capture time* rather than escape time; we stop computing as orbit of N_f when $|f(z)|$ becomes sufficiently small. For a family of f's one gets a *Newton–Mandelbrot* set by considering the critical orbits of N_f for each f; z is a critical point of N_f if $f''(z) = 0$.

4　Iterated function systems

In many cases, IFS attractors can be depicted on a computer screen with far less computing power (both in terms of hardware and software) than is needed for Mandelbrot and Julia sets. Fractint has an IFS option in which it reads data from an `ifs` file and generates the image in a manner I will describe below.

An *iterated function system* (IFS) in $\mathbb{R}^2$ (or more generally any complete metric space X) is a collection $\{F_1, \ldots, F_n\}$ of contraction mappings on $\mathbb{R}^2$. The main theorem of the subject says there is a unique nonempty compact subset $A \subset X$ (called the *attractor*) such that

$$A = F_1(A) \cup \cdots \cup F_n(A).$$

This is proved by considering a new metric space $\mathcal{H}(X)$ whose points are the nonempty compact subsets of X. To define the (Hausdorff) metric on $\mathcal{H}(X)$, first note the distance $d(K, L)$ between compact subsets K and L of X is the smallest number r such that each point in K is within r of some point in L and *vice versa*. If the metric space X is complete, so is $\mathcal{H}(X)$.

Given an IFS on X we define a mapping G of $\mathcal{H}(X)$ to itself by

$$G(K) = F_1(K) \cup \cdots \cup F_n(K).$$

If the F_i are contraction mappings, so is G: if each of the F_i shrinks distances by a factor of at most $s < 1$, then so does G. The contraction mapping theorem says that a contraction mapping on a complete metric space has a unique fixed point. Applying this observation to the contraction G on the complete metric space $\mathcal{H}(X)$ gives the existence and uniqueness of A.

Given a compact $K_0 \subset \mathbb{R}^2$, consider its orbit $\{K_i\}$ under G: $K_{i+1} = G(K_i)$. It is easily seen that

$$d(A, K_i) \leq \frac{s^i d(K_0, G(K_0))}{1 - s}.$$

Knowing s and $d(K_0, G(K_0))$, we can choose i so that this upper bound is less than the radius of a pixel. This means that for all practical purposes, A is the same as K_i. For a convenient choice of K_0 (such as a single point in the center of the screen), this gives a finite computation of A, called the *deterministic algorithm*.

The deterministic algorithm is easy to implement but not very fast, because it may have to compute each of the F_i on thousands of pixels in each iteration. A much faster method is the *random algorithm*. Here, choose a point $P_0 \in K_0$ and obtain P_{i+1} from P_i by applying *one* of the functions F_j, *chosen at random*. Then $P_i \in K_i$, so we know that for sufficiently large i, this point is within a pixel's radius of some point in A, and for such i we plot P_i on the screen. (Initial P_i need not be dropped if P_0 is taken to be the fixed point of one of the F_i. The fixed points of F_i belong to A, and so with this P_0 all the

P_i belong to A.) Each iteration requires computing just one function on one pixel, so the random algorithm is much faster than the deterministic algorithm.

The software (and much of Barnsley's book) deals with the case when the functions F_i are affine, i.e., of the form

$$F_i(x, y) = (a_i x + b_i y + e_i, c_i x + d_i y + f_i)$$

for $6n$ constants a_i through f_i. For the purpose of making the random choice, we assign a probability p_i to each F_i. Experience has shown that it is best to make p_i proportional to $|a_i d_i - b_i c_i|$, assigning a minimum positive probability when the determinant vanishes. FDesign does this automatically (after the user defines the $6n$ constants geometrically), but Fractint does not. With ths choice of p_i, the points produced by the random algorithm are evenly distributed over A.

It is instructive to see what happens when one does *not* choose the probabilities this way. In version 17.2 of Fractint (several years old) one can edit the ifs file and see the result immediately. Unfortunately, this feature is not present in later versions of the program.

5   Making fractal animations

In the summer of 1993, Jeffrey Lampert, a student in my first fractals course, hired on as a summer intern for me, and found animation software.

Fractint can save the images as gif files. A sequence of such files can be assembled into an fli or flc file, in effect a film, for which several display programs are available. Fractint also has a batch language (not the same as the formula language described below); it can respond to a series of instructions stored in a key file. (The Fractint package comes with a demonstration program that is a DOS batch file telling Fractint to read a file called demo.key. This batch language can be learned by examining that file.)

An animation typically consists of 100 or so incrementally varying images produced by Fractint. It is prohibitively tedious to make them all by hand, or to make them with a handwritten key file. Instead, in their favorite language the students write a program that will generate the desired file. If they are animating a sequence of IFS images, it is usually necessary to generate a specifically designed ifs file as well.

Once produced and saved in Fractint, the images have to be assembled into an animation file. The entire process of image production, saving, and assembly can take up to a minute of computing time per frame, depending on the speed of the computer and the resolution (usually 320×200 or 640×480) and the complexity of each image.

6   Some examples of fractint formulas

Fractint has a formula mode that is a simple programming language. It allows the user to define the function $f(z)$, the

initial point of each orbit, and the escape criterion. These typically vary with the pixel, and may also be controlled by two user defined complex parameters, p1 and p2. The formulas are stored in a text file (with the extension frm) read by Fractint, and the user may change the values of p1 and p2 each time a new image is created.

Fractint computes the orbit assigned to each pixel and colors it accordingly. When the program is written, the default coordinate system has screen corners at $\pm 2 \pm 1.5i$. Once the image has been produced, one has all of the usual Fractint options of zooming, reflecting, recoloring, etc.

Here are descriptions of some frm programs I have found useful.

(1) *Illustrating Lei's theorem.* (Lei (1990)) (This concerns the similarity between small regions of M and small regions of the corresponding Julia set.) Fractint can be programmed to show different images in different quadrants of the screen. Define characteristic functions $c_1(z)$, $c_2(z)$, $c_3(z)$, and $c_4(z)$, for each of the four quadrants. (The frm syntax includes the absolute value, real, and imaginary parts of a complex variable, so it is possible to make this definition.) Then by defining

$$f(z) = c_1(z) f_1(z) + c_2(z) f_2(z) + c_3(z) f_3(z) + c_4(z) f_4(z),$$

we are effectively looking at four different functions in the four quadrants of the screen. With the user defined parameter p1 for c, we can define f and the initial point of each orbit in such a way that the four screen quadrants are

- a neighboorood of c in M magnified 5 times,

- the same neighborhood magnified 100 times,

- the Julia set for c, and

- a neighborhood of c in the Julia set magnified 100 times.

(2) *Illustrating Feigenbaum points.* (A Feigenbaum point in M is a limit of a converging sequence of periodic points in which the period increases exponentially, rather than linearly as in the case of a preperiodic point.) This is similar to the previous example, but we show four successive magnifications of M around the point $c =$ p1, with the magnifications being powers of the number $\delta =$ p2. For complex values of δ, this 'magnification' includes a rotation. The best known example of a Feigenbaum point is $c = -1.401155\ldots$ (the limit point of the period doubling scenario), with $\delta = 4.669201\ldots$, the Feigenbaum number. Some other examples are listed in Milnor (1989). One can also use this program to illustrate self-similarity around preperiodic points, the simplest example being $c = -2$ with $\delta = 4$.

(3) *The* Curry, Garnett & Sullivan (1983) *experiment.* For the cubic function

$$f(z) = z^3 + (c - 1)z - c,$$

the authors study the Newton-Mandelbrot and Newton-Julia sets. The Newton function N_f has four critical points (where $f''(z) = 0$), three of which are roots of f. The orbit of the fourth critical point, $z = 0$, may or may not converge to a root. The values of c for which it fails to do so comprise the Newton-Mandlebrot set. The paper shows this set contains a small copy of the original M, centered at $c = .31 + 1.62i$, for which the critical orbit converges to a cycle of period 2. To see this Newton–Mandelbrot set with Fractint, set $c =$`pixel`, and iterate the function N_f starting at $z = 0$ and stopping when $|f(z)|$ gets small, say less than 10^{-6}. For a Newton-Julia set, take $c =$`p1`, $z =$`pixel` and iterate as before.

7   Conclusion

My experience with this course has convinced me that the study of fractals is a great way to draw students into mathematics. The image of a fractal on a computer screen has a fascination for both the specialist and the casual observer. Anyone curious about its properties and how it is produced quickly finds herself in deep mathematical waters. There is readily available software that makes it easy to illustrate lectures on fractals in real time. This subject is a motivational tool of which the mathematical community should make the most.

A Final Word

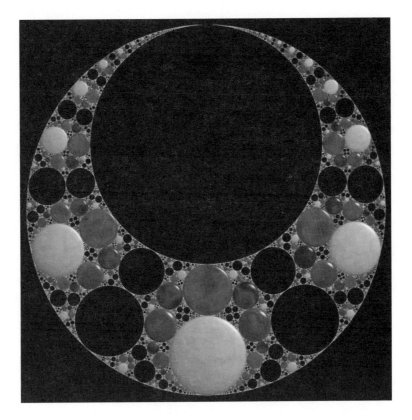

Chapter 17

The Fractal Ring from Art to Art through Mathematics, Finance, and the Sciences

Benoit B. Mandelbrot

In the beginning was the picture and the heartening story that is about to be told. Although it takes many odd turns, ultimately it ends with pictures.

A featherless biped does not become human until he conquers fire and spices and decorates bodies as well as homes and temples. For millennia, those decorative patterns become increasingly refined.

Some—homes, temples, pins and simple necklaces—help give birth to standard geometry.

Other decorative patterns are closer to nature but wait until around 1900 to enter mathematics. Great scholars, proclaiming themselves of a divine race and recognizing no ancestors, distill those patterns into monsters, shapes whose sole role is to free pure abstraction from the constraints of the real world and pictures.

Around 1960 the speaker reveals how those monsters can throw some light on the murkiness of financial markets.

In 1975 he publishes a book in which some esoteric mathematical monsters are named *fractals* and their fundamental features are—ironically—identified with those of many very familiar natural objects—including music. Because of past limitations, geometry and science had to leave those familiar objects aside and call them amorphous, that is, without specifiable form.

In 1982 the speaker publishes another book that builds on his earlier one and also attracts the mathematicians by using pictures to advance many conjectures, endlessly enchanting the specialists.

As the computer's graphic prowess grows, fractal images, once deemed to be solely utilitarian, are revealed to be spectacular. They enchant the students and specialists, who perceive them as decorative—even artistic.

Thus, having crossed and linked to one another several areas of disinterested or practical knowledge, and of feeling, the fractal ring finally closes: having begun long ago with art, it returns to its beginnings.

Appendices

Chapter 18

Panorama of Fractals and Their Uses

Michael Frame and Benoit B. Mandelbrot

When fractals were a brand new idea, and even when *The Fractal Geometry of Nature*, Mandelbrot (1982), was a new book being reviewed, futurists disagreed, as is their wont. A few envisioned a whole new field, but many did not. They found it more likely (as we did) that fractal ideas would spread far and wide among existing fields, influencing them to varying degrees but replacing or absorbing none. This second scenario is what is presently witnessed.

In the absence of a new field, no one is expected to produce a comprehensive treatise on fractals, but guidance among far divergent fields is indispensible. Some publications respond by including lists of diverse uses that tend to read like incomplete directories of a university's departments. They serve no positive purpose, antagonize some readers, and make us cringe. The panorama on which we settled instead could be viewed as a detailed wordbook, or lexicon, that orders alphabetically a wide mixture (inspired by the connectivity of the web) of entities of all sorts.

Our initial plan was to include a preliminary version of the Panorama in this volume, but a consequence of the remarkable range of fields that fractals reach is that even with some restraint, the Panorama grew too large. Yet we did not want to abandon this idea, so decided instead to introduce the Panorama as a web resource:

http://www.maa.org/Fractals/Panorama/
Welcome.html

In its current form, the Panorama is a growing web supplement to the text. Here we give a brief sketch of the development of the scope of the Panorama.

Initially we see three separate purposes.

1. The Panorama started modestly as a glorified index for the topics treated by the contributors to this book.

2. It developed by addressing itself specifically to the traditionally trained mathematicians who endeavor to help humanities and social science students to connect with mathematics as the science of patterns and changes. In that spirit, it expanded to incorporate examples that are classics in the fractalist community but known to few adults trained in mathematics.

3. Last, but not least, we view this as an early draft or work-in-progress. Initially we include mainly, but not exclusively, examples from the arts, humanities, and social sciences. These are generally less familiar to mathematicians, yet are very important to non-science students. The desire to post *something* while we are still living dictated many omissions in the initial offering of the Panorama.

A few of the entries are fairly long, to give a flavor of the ways in which this document can be a resource for teachers. Other entries are mere hints, included with the hope of inducing experts to submit contributions.

We emphasize the current Panorama is not intended to be an encyclopedia of all uses of fractals. Rather, it represents some we have found effective in the classroom. Many readers will note their favorite applications are missing. We encourage them to contact us (frame@math.yale.edu) about contributing to this website and to the eventual printed version of the Panorama.

As examples, here are two entries from the Panorama, one contributed by Nat Friedman, one by Peter Raedschelders.

Friedman. Fractals Bounding Negative Space.

Nathaniel Friedman is a mathematician specializing in ergodic theory, and a sculptor. In addition, since the early 1990s he has run a very successful annual conference on Art and Mathematics, and he is the director of the International Society of the Arts, Mathematics, and Architecture (http://math.albany.edu/isama/). One of Friedman's techniques involves fracturing slabs of granite, creating natural fractal boundaries. From some of these sculptures he makes prints. Here is his description of the process.

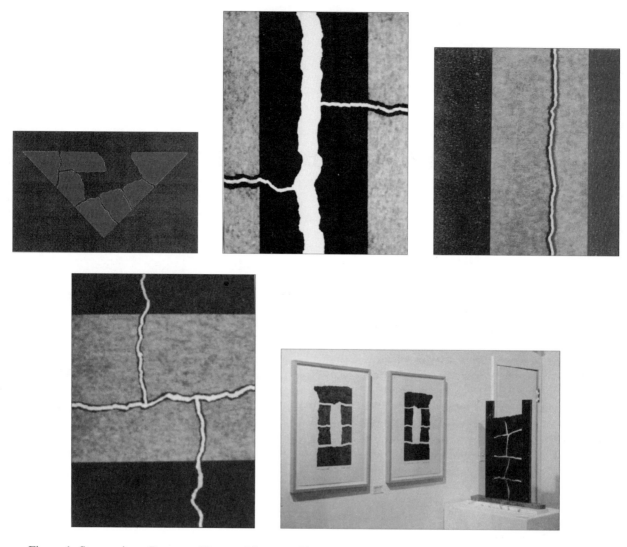

Figure 1: Stone prints: *Dancers*, *River and Streams*, *Flamenco*, and *Cliff Face*. A fractal sculpture and print pair.

Fractal Stone Prints

by Nathaniel Friedman, `artmath@csc.albany.edu`

For me, the major development in twentieth century sculpture was the opening up of the solid form and the creation of space as in the work of Barbara Hepworth and Henry Moore. I generally begin a sculpture by carving out a space. Like most sculptors, I accidently discovered that a shortcut to forming space was to break a stone and then separate the pieces to form a space. One then has the natural fractal geometry of the broken stone. Benoit Mandelbrot told me that, "the word fractal came from fracture, as in the edge of broken granite, rather then from fraction as some people think." This statement definitely influenced me since I had been working with broken granite and was led to a technique for making prints. I begin with a planar shape of one-inch thick polished granite. On the rough side I draw a pattern of straight lines along which I tap with a chisel to split the granite along the lines. This results in a fractal division of the planar shape into several pieces. I separate them to form space between pieces and I may also remove some pieces to form spaces. Black ink is rolled on the polished surface of the stones and a sheet of thin Japanese paper is placed on the inked stones. When pressure is applied with a barren, the ink permeates the paper. The lower surface of the paper on the stones has a solid black image, which was the original intent. It came as a surprise that the ink permeating the paper produced an interesting grey-black image on the upper surface that is visible when the print is being made. By applying pressure with a burnishing tool along the broken edges, these fractal edges show up as black on the upper surface due to the ink permeating the upper surface of the paper. Thus, one actually obtains a two-sided print, where the lower side is black and the upper side is a grey-black monoprint controlled by the pressure of the barren and the burnishing tool. One can also fold one side onto the other to obtain an image combining both sides. One can also use colored inks, which has opened up many more possibilities.

Figure 1 shows four examples, and a picture of a sculpture and two prints made from it.

I consider the fractal geometry of broken granite to be a visually exciting geometry in contrast to straight line Euclidean geometry. I have control over the pattern of straight lines that I draw on the back of the stone along which I tap with a chisel to break the stone. However, I have no control over the way the stone breaks to form natural fractals, which introduces a certain randomness. In a print such as *River and Streams*, I call attention to the fact that the fractal geometry of broken granite and the fractal geometry of rivers and streams is the same. In the print the river is represented by the wider spacing of the stones and the streams correspond to the narrow spacing. Part of the process is selecting appropriate spacings. In some prints I just like working with the natural fractal geometry for its own sake without any underlying meaning.

I have also made some abstract fractal torsos that are done by breaking a stone, removing some inner pieces, and then reassembling the remaining pieces. Here the fractal geometry reflects the psychological process of coming apart and putting yourself back together. The inner space may have a positive or negative interpretation. It could represent getting rid of something that bothered you or it could represent the loss of someone. In any case, only fractal geometry can convey this psychological process of breaking apart. Here is where fractal geometry reflects nature in psychology.

Tilings: Raedschelders'[1] fractals following Escher

The graphic design work of Maurits Escher is familiar to most people having interest in both art and geometry. Many of Escher's best-known works involve turning tilings of the plane (checkerboards, for example) into tilings by shapes of animals, or horsemen, or mythical figures. But Escher was disturbed by problems of representing infinite spaces within finite bounds. In one approach to this problem, he dropped the familiar geometry of Euclid and instead constructed his tilings using hyperbolic geometry. See Coxeter (1979). *Circle Limit III* is one of his clearest examples.

Escher died before fractals were well known. A natural speculation is what Escher would have done if he had been familiar with fractals. Enclosing infinite complexity within a bounded space is certainly something one can do with fractals. Peter Raedschelders did such an experiment, and reports on it here. Raedschelders is an engineer interested in art and mathematics. He has made about 40 prints, mostly inspired by, but often going further than, themes developed by Escher. His work has been shown in numerous exhibitions, and featured in books, including the covers of two mathematics texts.

---

[1]Thanks to Brooke Crowder for bringing Peter Raedschelders' work to our attention.

Making the Fractal Tessellation Print "Butterflies I"

by Peter Raedschelders,
`peter.raedschelders@planetinternet.be,`
`http://home.planetinternet.be/~praedsch`

As an admirer of the work of M.C. Escher, the famous Dutch artist who made several beautiful tessellations, I tried to find out how tessellations could be made.

There are several methods which can be used for making simple regular tessellations. Today many of these methods can be found in books or on the internet. A method which you will not find in books is just to look at the prints of Escher and do a little bit of trial and error. This method is great fun, and that is what I did.

Of course one must start with the easy prints, regular tessellations that cover the complete plane. See Figure 2.

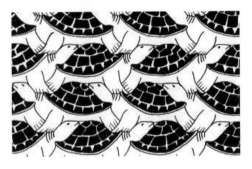

Figure 2: Turtles: a regular tessellation.

But covering the complete plane is a lot of work, and very soon, one is no longer satisfied by this type of tilings. Again I looked at the work of the master and I got fascinated by tessellations with limits.

Escher showed us that it was possible to make tessellations with border-limits, so it was no longer necessary to cover the complete plane.

He used hyperbolic geometry for his woodcuts "Circle-limits" (pages 429, 432, and 434 of Locher (1981)). Escher made several other prints with limits, so I tried to find out if there are other types of border-limits.

The first trial, the left side of Figure 3, was a tessellation with one border-limit.

Now we use a limit but we still need the complete plane.

Afterwards we used a very simple construction to make a tessellation with a real border-limit. Take a square, divide it by drawing a diagonal so you get two isosceles triangles. On each of the shortest sides place smaller isosceles triangles. If we continue this process we get a tessellation of smaller and smaller triangles and a border-limit. After deforming the straight lines into curves, (and with a little bit of imagination) seals appear. See the right side of Figure 3.

What we did with a square we can also try to do with a regular hexagon. Divide the hexagon and use smaller ver-

Figure 3: Left: Turtles with a single border-limit. Right: Seals enclosed by a border-limit.

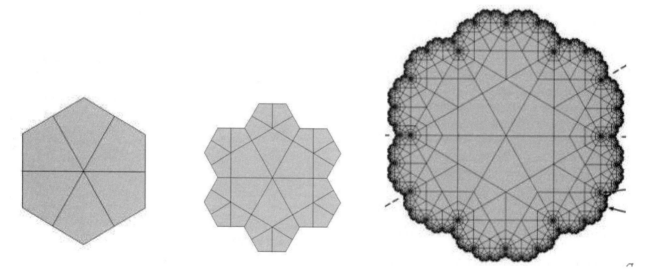

Figure 4: Left and center: the first two steps. Right: the basis drawing (Robert Fathauer, with permission).

sions of the pieces to build some kind of tessellation with a border-limit. The result is Figure 4. Dr. Robert Fathauer also discovered this fractal together with several others, all can be found in

http://www.tessellations.com/encyclopedia.html

An interesting thing about this drawing is that it has not only a border-limit but also several limits within the figure.

Now the next problem was to find some kind of animal or other recognizable figure that fits into this tessellation. We had segments which had all the same shape but that shape had a good correspondence to no animal. It is always nice to have only one recognizable figure in a tessellation, but sometimes this makes it also really hard to imagine such a figure.

Therefore, it was considered better to take two different recognizable figures. See Figure 5.

Figure 5: Left: Butterfly type 1. Right: Butterfly type 2.

When two segments are joined together in two different ways, two different kinds of butterflies could be drawn. In the center another limit was created, although this part of the print could also be filled by three butterflies of type 2.

Now the complete basis drawing was filled in with hundreds of butterflies, all drawn by hand. (At that time I didn't have a computer.) See Figure 6.

Figure 6: "Butterflies I."

We admit that the fitting is not perfect, there are some gaps between the butterflies, but we believe that now the print looks better than with a perfect fitting.

The print was made about 1984 and at that time fractals were unknown to me. For me it was just a good looking print with several limits, just another print to show that mathematics can not only be very interesting but also be beautiful.

Figure 7 shows two more examples of Robert Fathauer's extensive catalog of fractal tilings.

As a final enticement to consult the website, Figure 8 shows one of Ken Musgrave's artificial landscapes from the *Mountains* entry of the Panorama.

Conclusion

Simple ideas that fundamentally alter our perception of the world are very rare. Overenthusiastic zealots of fractal geometry believe already there is sufficient evidence to guarantee that role; rabid critics assert fractal geometry is a passing fad, soon to disappear from sight. (How long must we wait before a passing fad passes? Twenty years seems a bit long.) We believe fractal geometry will make fundamental changes in science and far beyond, but are waiting for the evidence to accumulate. One of the roles of this Panorama is to share some evidence we have already seen. Another role, more important in our eyes, is to invite you to share your examples with us. In this way, we hope this Panorama indeed will grow into *A Panoramic Encyclopedia of Fractals and Their Uses*.

We're waiting to hear from you.

Figure 7: Two fractal tilings.

Figure 8: A fractal landscape, in fact, part of an artificial planet.

Chapter 19

Reports of Some Field Experiences

Together with Nial Neger we have run summer workshops for high school and college teachers, mostly from Connecticut. A total of about 90 teachers attended in 2000 and 2001. Many of those who attended the 2000 workshop incorporated some aspects of fractal geometry in their 2000–01 courses, and we expect more from the second group.

We thought it appropriate to end this volume with brief reports from a few graduates from our workshops.

Carol Ann Dobek   Simsbury High School, Simsbury, CT

I had my AP (BC) calculus class do research and an oral report about fractals in any discipline. The student projects, from basic fractals to various applications, went beyond my expectations. After the initial research, most students became very excited about their work. After the oral presentations, some of the students were motivated to do more investigations into other areas of fractals.

Patricia McGrath   Maloney High School, Meridan, CT

A challenge to explore fractals augments initiative, creativity, and teamwork in students. At the end of the school year, seniors were asked to collect fractal curiosities and to give PowerPoint presentations to class. Compacted into a couple of days, student work exhibited evidence of competent interest, research, PowerPoint planning, and enthusiasm for extended explorations into art, music, physiology, scenery design, etc. Also, they were exposed to topics of interest to whet their intellectual appetites to discover and explore even beyond their high school graduation.

Phyllis Pruzinsky   Trumbull High School, Trumbull, CT

I used fractals in my Pre-calculus classes to reinforce the four main transformations of functions: reflections, dilations, rotations, and translations. The students first learn these transformations for functions (linear, quadratic, cubic, absolute value) with which they are familiar. Then, throughout the year, they have to use these transformations with new functions (trigonometric, exponential, logarithmic).

After teaching the transformations, I took three days to have students use these transformations to find the IFS rules to generate several fractals. The first day included a general introduction to fractals and the notation needed to write the IFS rules. I had students use manipulative pieces to help see the transformations for two or three simple fractals.

On the second day I began with a demonstration on the overhead of three fractals involving 3 or 4 transformations. Then students were given several fractals and the necessary manipulative pieces and asked to write the IFS rules for them.

The third day was spent in the computer lab where students were able to test their IFS rules, using Nial Neger's IFS program. They were given time to correct any errors and go on to more difficult fractals.

The students truly enjoyed this puzzle-like method of reinforcing transformations of functions. They challenged each other to see who could generate the fractal first. The excitement level in the classroom was tremendous! Their original reluctance to learn this new way-out topic turned into enthusiasm like they would have playing a fun and challenging computer game.

Ann Robertson   Connecticut College, New London, CT

The Fractal Geometry and Chaos units in my *Introduction To Mathematical Thought* course are stronger because of the Fractal Geometry workshops and your web site! Furthermore in the future, I plan to expand the fractals and chaos material to a full semester offering in our *Introduction To Mathematical Thought* series. Material on Iterated Function Systems, box counting dimension, and basins of attraction will be most relevant. Also, the software presented in your workshops will assist in bringing a hands-on component to the course. One text for the course material will most certainly be your web site. The numerous applications there will indeed help me organize that section of the course.

I will be teaching a one semester *Introduction to Calculus* in the fall. My Newton's Method lab using Mathematica will be enlarged using the basins of attraction material from the second workshop. My original reference for the lab was from *Learning By Discovery*, MAA Notes, Number 27, pg. 49. One of the problems in the lab deals with sensitivity to initial conditions of the function $f(x) = x^3 - x$ and so all the material on basins of attraction is most relevant.

Previously, conductor Michael Adelson and I combined our advisees (potential music or math majors) for two tutorial sessions: (1) Introduction to Fractal Geometry and (2) Fractals in Beethoven's *Seventh Symphony*. The material was relevant to both groups and the interdisciplinary nature of our talks made the sessions interesting for the students.

Rita Robbins and Dianne Foster from Bennie Dover Jackson Middle School and I worked as one team combining Connecticut College and BDJMS's resources to introduce Fractal Geometry into the eighth grade curriculum. This was instituted under a Lucent Technologies Teach and Learn Partnership grant. We met throughout the year at the college and then I visited eighth grade classes. The classes were exciting and the students were eager to learn about the Sierpinski gasket, etc. I had fun being an observer. Then in May we had the entire eighth grade come to the college for a presentation on Fractal Geometry and its applications. To keep their attention, we used lecture material (PowerPoint), video clips, a demonstration (your finger painting one), Internet sites, including yours, and a five minute writing exercise. (Used not only for closure each day, but for follow-up for the writing instructors and for assessment purposes.) Our program for each of the two groups was two hours long (probably too long but the kids were great.) They wrote answers to such questions as (1) What one new thing did you learn today? (2) What did you like the best? and (3) Do you have any questions? I can say we were all stunned and thrilled by their responses. "I learned about self similarity," "I never understood infinity until I saw the Mandelbrot set," "I liked the zooming in the best," "I'll look at clouds now," were just a few of their very interesting responses. I ended each session by sharing with the students what they had written. Wow, look what you started! In a small way, I have tried to give back a bit of what I have gained from the workshops.

We had 40 to 45 eighth graders the first day and 3 teachers and over 70 students with 8 teachers on the second! I do remember a rather funny incident that you will enjoy. On the second day the young people were squirmy so I gave them a bathroom break early on. One instructor then came up to me and said "Don't do that! No more breaks!" I immediately agreed thinking she must know more than I. I must have looked puzzled as she followed up by saying "I don't know anything about this (Fractal Geometry). I want to learn!"

Under the same grant program, Jeffrey Wolfson, an eighth grade art instructor and I met to highlight the mathematics, including fractals, that is present in art. I visited a couple of his classes and attended a most wonderful art show. We intend to formally introduce a *Math in the Art Curriculum* next year including having the students experiment with Jackson Pollock's drip method. (Could get messy so we will probably do that last!) Jeff is both very creative and adventuresome. His students were working on pottery and symmetry the day I attended. I was truly impressed.

The department introduced a MAT495 topics course for seniors majoring in math this past year. The idea was for profs to share their research with the students and for students to be exposed to current problems in mathematics. Since I am a math education type, I decided on a topic involving Fractal Geometry and Chaos. (Somewhat similar to the talk I gave in your April session.) The students' assignment was to read Stoppard's *Arcadia*, see the movie *Pollock* and listen to Beethoven's *Seventh Symphony* before the lecture. As an optional assignment (next time it will not be optional) I gave them Nial's box counting exercise. Students not only expressed interest in the material but enjoyed the varied approach of linking fractals and the arts. Journal entries, followup questions, and evaluations were all very positive!

Stephen Smith   Enfield High School, Enfield, CT

No longer can I recall when it was that I first caught the fractal bug. I know that it was more than seven years ago and it was visual (that beetle-like black thing). The Mandelbrot set entered my life to stay.

However, I was frustrated during the next few years by my inability to grasp the mathematics to an extent sufficient to share it with my high school students. I saw the video *The Colors of Infinity*, loved it, and show it to my classes. I showed them on a limited basis how to compute the Mandelbrot set. Yet the concept was vague. Then last spring I heard about a Fractal Geometry workshop at Yale and immediately signed up. In 2000, along with about 45 other teachers, I attended Michael Frame's Fractal Geometry Workshop.

Awestruck, I sat as the incredible world of fractals unfolded before me! I had never imagined the width of the spectrum of topics that fractal ideas encompasses. Michael Frame and Nial Neger do a wonderful (wonder-filled!) job. To listen to Benoit Mandelbrot himself (who gave two lecture-discussions) is an unimaginable treat! Michael's web-based lecture style defies comparison. The web pages are superb. The activities are great.

During the subsequent school year I was able to sprinkle my classes (*Preparation for Algebra*, *Algebra II*, and *PreCalculus*) with fractal concepts. By the end of the year I had assembled a unit[1] of Cellular Automata and Fractal Dimensions that I used with my lower-level students. I plan to incorporate it into *Algebra II* and *PreCalculus*.

In April 2001 I attended a one-day reunion workshop and we have just completed a three-day Workshop II. My knowledge of fractals has increased and I now have several more activities to share with my students and colleagues.

I have nothing but praise for these workshops and the effect they have had on me and my ability to communicate the concepts and potential of Fractal Geometry. I highly recommend them to anyone who wants to be amazed.

---

[1] Stephen shared this material with us during the 2001 Workshop II. We incorporated it into the 2001 Workshop I. It drew very positive responses, and several teachers immediately said they would use it in their 2001–2002 classes.

A Guide to the Topics

The first digit of each entry refers to a chapter number, the following digits to sections.

References

References

Allman, W. (1990) *Apprentices of Wonder: Inside the Neural Network Revolution*, Bantum Press, New York.

Anton, H., and Rorres, C. (1991) *Linear Algebra*, Addison Wesley.

Argyros, A. J. (1991) *A Blessed Rage for Order: Deconstruction, Evolution, and Chaos*, University of Michigan Press, Ann Arbor, MI.

Banks, J., Brooks, J., Cairns, G., Davis, G., and Stacey, P. (1992) "On Devaney's Definition of Chaos," *Amer. Math. Monthly* **99** 332–334.

Barnsley, M. (1988) *Fractals Everywhere*, Academic Press, San Diego.

Barnsley, M. (1993)*Fractals Everywhere* , Academic Press, San Diego. (2nd Ed.)

Barnsley, M. F. and Hurd, L. B. (1993) *Fractal Image Compression*, AK Peters Ltd. MA, Wellesley, MA.

Bassingthwaite, J. B., Liebovitch, L. S., and West, B. J. (1994) *Fractal Physiology,* Oxford Univ. Press, Oxford.

Berkey, D., and Blanchard, P. (1992) *Calculus*, Saunders College Publishing, Orlando.

Berreby, D. (1993) "Chaos hits Wall Street," *Discover*, 76–84, March.

Bishop, C., and Jones, P. (1997) "Hausdorff dimension and Kleinian groups," *Acta Mathematica* **179** 1–39.

Briggs, J., Peat, F. (1989) *Turbulent Mirror: An Illustrated Guide to Chaos Theory and the Science of Wonder*, Harper and Row, New York.

Bunde, A., and Havlin, S. (Eds.) (1991) *Fractals and Disordered Systems,* Springer-Verlag, Berlin and New York.

Camp, D., Chiaverina, C., and Senior, T. (1999) "Festive fractals," *The Physics Teacher*, 532.

Carlson, C. (1991) "Spatial distribution of ore deposits," *Geology* **19** 111–114.

Coxeter, C. A. (1979) "The non-Euclidean symmetry of Escher's picure 'Circle Limit III'," *Leonardo* **12** 19–25.

Crichton, M. (1990) *Jurassic Park*, Alfred A. Knopf, New York.

Crownover, R. (1995) *Fractals and Chaos,* Jones and Bartlett, Boston.

Cuomo, K. M., Oppenheim, A. V., and Strogatz, S. H. (1993) "Synchronization of Lorenz-based chaotic circuits with applications in communications," *IEEE Transactions: Analog and Digital Signal Processing* **40** 626–633.

Curry, J. H., Garnett, L., and Sullivan, D. (1983) "On the iteration of a rational function: computer experiments with Newton's method," *Commun. Math. Phys.* **91** 267–277.

Davis, D. M. (1993) *The Nature and Power of Mathematics*, Princeton Univ. Press, Princeton.

Devaney, R. L. (1989) *Chaos, Fractals, and Dynamics: Computer Experiments in Mathematics*, Addison-Wesley, Menlo Park.

Devaney, R. L. (1989) *An Introduction to Chaotic Dynamical Systems*, Second Edition, Addison-Wesley Co., Menlo Park.

Devaney, R. L. (1990) *Chaos, Fractals and Dynamical Systems: Computer Experiments in Mathematics*, Addison-Wesley, Menlo Park.

Devaney, R. L. (1991) "The Orbit Diagram and the Mandelbrot Set," *The College Mathematics Journal* **22** 23–38.

Devaney, R. (1992) *A First Course in Chaotic Dynamical Systems: Theory and Experiment*, Addison-Wesley, Menlo Park.

Devaney, R. L. (ed.) (1994) *Complex Analytic Dynamics: The Mathematics Behind the Mandelbrot and Julia Sets*, American Mathematical Society, Providence.

Devaney, R. L. (1995) "Explorations in the chaos club," *Focus.* **15** No. 3, 8–9.

Devaney, R. L. (1999) "The Mandelbrot Set and the Farey Tree," *Amer. Math. Monthly* **106** 289–302.

Devaney, R. L. and Keen, L., (eds.) (1989) *Chaos and Fractals: The Mathematics Behind the Computer Graphics*, Proc. Sympos. Applied Math., American Mathematical Society, Providence, 2nd Rev Ed.

Devlin, K. (1989) "Chaos," *Mathematical Intelligencer*, **11** No. 3, 3.

Dewdney, A. K. (1985) "Computer Recreations," *Scientific American* **284** 16–32.

Ditto, W., Rauseo, S., and Spano, M. (1990) "Experimental control of chaos," *Physical Review Letters* **65** 3211–3214.

Douady, A. (1986) "Algorithms for computing angles in the Mandelbrot set," *Chaotic Dynamics and Fractals* (M. F. Barnsley and S. G. Demko, eds.), Academic Press, pages 155–168.

Douglas, R. (1989) "Chaos," *Mathematical Intelligencer*, **11** No. 3, 3–4.

Edgar, G. A. (ed.) (1993) *Classics on Fractals*, Addison-Wesley, Menlo Park.

Edgerton, S. (1992) *The Heritage of Giotto's Geometry: Art and Science on the Eve of the Scientific Revolution*, Cornell University Press.

Evertsz, C. J. G., H.-O. Peitgen, and R. F. Voss (eds.) (1996) *Fractal Geometry: The Mandelbrot Festricht,* Special Volume in Honor of Benoît B. Mandelbrot's 70th Birthday, World Scientific, Singapore.

Falconer, K. (1990) *Fractal Geometry: Mathematical Foundations and Applications*, John Wiley, Chichester.

Farruggia, C., Lawrence, M., and Waterhouse, B. (1996) "The Elimination of a Family of Periodic Parity Vectors in the $3x + 1$ Problem," *Pi Mu Epsilon Journal* **10** 275–280.

Feder, J. (1988) *Fractals,* Plenum Press, New York, 1988.

Feder, J., and Aharony, A. (eds.) (1990) *"Fractals in Physics: Essays in Honor of B. B. Mandelbrot,"* North Holland, Elsevier.

Fermi, E. (1965) *Collected Papers II*, University of Chicago Press.

Fermi, E., Pasta, J., and Ulam, S. (1955) "Studies of Non Linear Problems," LA-1940 paper no. 266, Los Alamos.

Fraboni, M., "Conjugacy and the $3x + 1$ Conjecture," submitted.

Frame, M., and Lanski, J. (1999) "When is a recurrent IFS attractor a standard IFS attractor?," *Fractals*, **7**, 257–266.

Franks, J. (1989) "Chaos: Making a New Science by James Gleick," *Mathematical Intelligencer*, **11** No. 1.

Franks, J. (1989) "Comments on the Responses to my Review of Chaos," *Mathematical Intelligencer*, **11** No. 3, 12–13.

Fusaro, M. (1997) "A Visual Representation of Sequence Space," *Pi Mu Epsilon Journal* **10** 466–481.

Gardner, M. (1992) *Fractal Music, Hypercards, and More*, W.H. Freeman, New York.

Garfinkel, A., Spano, M., Ditto, W., and Weiss, J. (1992) "Controlling cardiac chaos," *Science* **257** 1230–1235.

Ghyka, M. (1977) *The Geometry of Art and Life*, Dover, New York.

Gilmore, C. (1993) "A new approach to testing for chaos, with applications in finance and economics," *International Journal of Bifurcation and Chaos* **3** 581–587.

Gleick, J. (1987) *Chaos: Making a New Science,* Viking Penguin, New York.

Gleick, J. (1989) "James Gleick Replies," *Mathematical Intelligencer*, **11** No. 3, 8–9.

Goldberger, A., Rigney, D., and West, B. (1990) "Chaos and fractals in human physiology," *Scientific American*, February, 42–49.

Goodman, A. (1974) *Analytic Geometry and the Calculus*, Macmillan.

Griffith, C., Lapidus, M. (1997) "Computer graphics and the eigenfunctions for the Koch snowflake drum," in *Progress in Inverse Spectral Geometry*, Birkhaüser-Verlag, Basel and Boston, pages 95–109.

Hadamard, J. (1898) *Leçons de Géométrie élémentaire (Géométrie plane)*, Armand Colin, Paris.

Hardy, G. (1940) *A Mathematician's Apology*, Cambridge University Press, Cambridge.

Hayes, S., Grebogi, C., Ott, E., Mark, A. (1994) "Experimental control of chaos for communication," *Physical Review Letters* **73** 1781–1784.

Hayles, N. K. (1991) "Complex dynamics in literature and science," in *Chaos and Order: Complex Dynamics in Literature and Science*, N.K. Hayles, ed., University of Chicago Press, Chicago, pages 1–33.

He, C., Lapidus, M. (1996) "Generalized Minkowski content and the vibrations of fractal drums and strings," *Math. Research Letters* **3** 31–40.

He, C., Lapidus, M. (1997) "Generalized Minkowski content, spectrum of fractal drums, fractal strings and the Riemann zeta-function," *Memoirs Amer. Math. Soc.* No. 608, **127** 1–97.

Henderson, L. (1983) *The Fourth Dimension and Non-Euclidean Geometry in Modern Art*, Princeton University Press.

Hilbert, D., Cohn-Vossen, S. (1952) *Geometry and the Imagination*, Chelsea, New York. English translation of *Anshauliche Geometrie* (1932).

Hirsh, M. (1989) "Chaos, Rigor and Hype," *Mathematical Intelligencer*, **11** No. 3, 6–8.

Hocking, J. and Young, G. (1961) *Topology*, Addison-Wesley, Reading.

Hsü, K., Hsü, A. (1990) "Fractal geometry of music," *Proc. Natl. Acad. Sci. USA* **87** 938–941.

Hsü, K., Hsü, A. (1991) "Self-similarity of the '$1/f$' noise' called music," *Proc. Natl. Acad. Sci. USA* **88** 3507–3509.

Hunt, E. (1991) "Stabilizing high-period orbits in a chaotic system: the diode resonator," *Physical Review Letters* **67** 1953–1955.

Hutchinson, J. (1981) "Fractals and self-similarity," *Indiana University Journal of Mathematics* **30** 713–747.

Janson, H. (1991) *History of Art*, 4th ed., Abrams.

Jeffrey, H. (1992) "Chaos game visualization of sequences," *Computers & Graphics* **16** 25–33.

Joseph, J., "A Chaotic Extension of the $3x + 1$ Function to $Z_2[i]$," submitted.

Kaye, B. K. (1989) *A Random Walk Through Fractal Dimensions*, VCH Publishers, New York.

Kelly, E. (1992) *Ellsworth Kelly: the Years in France 1948–1954*, National Gallery of Art, Washington D. C.

Kennedy, J. Yorke, J. (1991) "Basins of Wada," *Physica D*, **5** 213–225.

Kern, A. (1997) "IFS variations of the Sierpinski gasket: predicting connectedness with line preservation," term project, *Frctal Geometry*, Union College.

Kigami, J., Lapidus, M. (1993) "Weyl's problem for the spectral distribution of Laplacians on p.c.f. self-similar fractals," *Commun. Math. Phys.* **158** 93–125.

Lapidus, M. (1991) "Fractal drum, inverse spectral problems for elliptic operators and a partial resolution of the Weyl-Berry conjecture," *Trans. Amer. Math. Soc.* **325** 465–529.

Lapidus, M. (1992) "Spectral and fractal geometry: From the Weyl-Berry conjecture for the vibrations of fractal drums to the Riemann zeta-function," in *Differential Equations and Mathematical Physics,* (C. Bennewitz, ed.), Academic Press, New York, pages 151–182.

Lapidus, M. (1993) "Vibrations of fractal drums, the Riemann hypothesis, waves in fractal media, and the Weyl-Berry conjecture," in *Ordinary and Partial Differential Equations,* (B. Sleeman and R. J. Jarvis, eds.) Longman, London, pages 126–209.

Lapidus, M. (1994) "Analysis on fractals, Laplacians on self-similar sets, noncommutative geometry and spectral dimensions," *Topological Methods in Nonlinear Analysis* **4** 137–195.

Lapidus, M. (1995) "Fractals and vibrations: Can you hear the shape of a fractal drum?," in *Fractal Geometry and Analysis* World Scientific, Singapore, pages 321–332.

Lapidus, M. (1997) "Towards a noncommutative fractal geometry? Laplacians and volume measures on fractals," *Contemporary Mathematics*, Amer. Math. Soc. **208** 211–252.

Lapidus, M., Maier, H. (1995) "The Riemann hypothesis and inverse spectral problems for fractal strings," *J. London Math. Soc.* **52** 15–35.

Lapidus, M., Neuberger, J., Renka, R., and Griffith, C. (1996) "Snowflake harmonics and computer graphics: Numerical computation of spectra on fractal domains," *Intern. J. Bifurcation & Chaos* **7** 1185–1210.

Lapidus, M., Pang, M. (1995) "Eigenfunctions of the Koch snowflake domain," *Commun. Math. Phys.* **172** 359–376.

Lapidus, M. L., Pomerance, C. (1993) "The Riemann zeta-function and the one-dimensional Weyl-Berry conjecture for fractal drums," *Proc. London Math. Soc.* **66** 41–69.

Lapidus, M., Pomerance, C. (1996) "Counterexamples to the modified Weyl-Berry conjecture on fractal drums," *Math. Proc. Cambridge Philos. Soc.* **119** 167–178.

Lapidus, M., van Frankenhuysen, M. (1999a) "Complex dimensions of fractal strings and oscillatory phenomena in frac-tal geometry and arithmetic," in *Spectral Problems in Geometry and Arithmetic,* (T. Branson, ed.), *Contemporary Mathematics* **237** 87–105.

Lapidus, M., van Frankenhuysen, M. (1999b) *Fractal Geometry and Number Theory, Complex Dimensions of Fractal Strings and Zeros of Zeta-Functions.* Birkhäuser, Boston.

Lauweier, H. (1991) *Fractals: Endlessly Repeating Geometrical Patterns*, Princeton University Press, Princeton.

Lawler, G., Werner, W., Schramm, O. (2000) "Values of Brownian intersection exponents I: Half-plane exponents," "Values of Brownian intersection exponents II: Plane exponents," and "Values of Brownian intersection exponents III: Two-sided exponents," to appear in *Acta Mathematica*. Preprints are available at

`http://www.math.u-psud.fr/~werner/pre.html`

Lei, T. (1990) "Similarity between the Mandelbrot set and Julia sets," *Commun. Math. Phys.* **134** 567–616.

Lightman, A. (1994) *Einstein's Dreams*, Warner, New York.

Locher, J. (1981) *M. C. Escher*, Meulenhoff, Amsterdam.

Mandelbrot, B. (1963) "The variation of certain speculative prices," *The Journal of Business* **36** 394–419.

Mandelbrot, B. (1967) "How long is the coast of Britain? Statistical self-similarity and fractional dimension," *Science* **156** 636–638.

Mandelbrot, B. (1975) *Les objets fractals: forme, hasard et dimension*, Flammarion, Paris.

Mandelbrot, B. (1980) "Fractal aspects of the iteration $z \to \lambda z(1 - z)$ for complex λ and z," *Annals of the New York Academy of Sciences* **57** 249–259.

Mandelbrot, B. (1982) *The Fractal Geometry of Nature*, W. H. Freeman, San Francisco.

Mandelbrot, B. (1983) "Self-inverse fractals osculated by sigma-discs and the limit sets of inversion groups," *Mathematical Intelligencer* **5** 9–17.

Mandelbrot, B. (1989) "Fractals and an art for the sake of science," *Leonardo, Supplemental Issue Computer Art in Context*, 21–24.

Mandelbrot, B. (1992) "Fractals, the Computer, and Mathematics Education," closing address, International Congress on Mathematics Education, ICME-7, Quebec City. Reprinted (with modifications) in this volume.

Mandelbrot, B. (1995) "Introduction to fractal sums of pulses," *Lévy Flights and Related Topics in Physics*, M. Shlesinger, G. Zaslavsky, U. Frisch, eds, Springer-Verlag, Berlin.

Mandelbrot, B. (1997) *Fractals and Scaling in Finance, Discontinuity, Concentration, Risk, Selecta Vol E*, Springer-Verlag, New York.

Mandelbrot, B. (1999) *Multifractals and 1/f Noise. Wild Self-Affinity in Physics, Selecta Vol N,* Springer-Verlag, New York.

Mandelbrot, B. (2002) *Gaussian Self-Affinity and Fractals: Globality, the Earth, 1/f and R/S, Selecta Vol H,* Springer-Verlag, New York.

Mandelbrot, B., Passoja, D., Paullay, A. (1984) "The fractal character of the fracture surfaces of metals," *Nature* **308** 721–722.

Mandelbrot, B., Stauffer, D. (1994) "Antipodal correlations and the texture (fractal lacunarity) in critical percolation clusters," *Journal of Physics A* **27** L237–L242.

Mandelbrot, B., Vespignani, A., Kaufman, H. (1995) "Crosscut analysis of large radial DLA: departures from self-similarity and lacunarity effects," *Europhysics Letters* **32** 199–204.

May, R. (1976) "Simple mathematical models with very complicated dynamics," *Nature* **261** 459–467.

McCorduck, P. (1991) *Aaron's Code: Meta-Art, Artificial Intelligence, and the Work of Harold Cohen,* Freeman.

Milne, B. (1991) "Lessons from applying fractal models to landscape patterns," in *Quantitative Methods in Landscape Ecology,* M. Turner and R. Gardner, eds., Springer-Verlag, New York.

Monks, K., "How to Create a Successful Undergraduate Student Research Program in Mathematics in Your Spare Time Starting with No Cash," preprint.

Musgrave, K., Mandelbrot, B. (1991) "The art of fractal landscapes," *IBM Journal of Research and Development* **35** 535–540.

Nonnenmacher, T., Losa, A., Weibel, E. , eds. (1994) *Fractals in Biology and Medicine,* Birkhäuser-Verlag, Basel and Boston.

Nusse, H., Yorke, J. (1996) "Basins of attraction," *Science* **271** 1376–1380.

Parris, R. (1990) "The Root-Finding Route to Chaos," *The College Mathematics Journal,* **22** 42–55.

Passeron, R. (1978) *Phaidon Encyclopedia of Surrealism,* Phaidon Press.

Patnode, B. (1995) "An Exhibition of Prints," Atrium Gallery, Union College, April.

Peak, D., Frame, M. (1994) *Chaos Under Control: The Art and Science of Complexity,* Freeman, New York.

Pearse, R. (1998) "Universality of the Sierpinski Carpet," Honors Undergraduate Thesis, Mathematics, University of California, Riverside, Calif.

Peitgen, H.-O., Jurgens, H., Saupe, D. (1991) *Fractals for the Classroom. Strategic Activities, Volume One,* Springer-Verlag, New York.

Peitgen, H.-O., Jürgens, H., Saupe, D. (1992a) *Chaos and Fractals: New Frontiers of Science,* Springer-Verlag, New York.

Peitgen, H.-O., Jurgens, H., Saupe, D. (1992b) *Fractals for the Classroom, Part One. Introduction to Fractals and Chaos,* Springer-Verlag, New York.

Peitgen, H.-O., Jurgens, H., Saupe, D. (1992c) *Fractals for the Classroom, Part Two. Introduction to Fractals and Chaos,* Springer-Verlag, New York.

Peitgen, H.-O., Jurgens, H., Saupe, D. (1992d) *Fractals for the Classroom. Strategic Activities, Volume Two,* Springer-Verlag, New York.

Peitgen, H.-O., Richter, P. (1986) *The Beauty of Fractals: Images of Complex Dynamical Systems,* Springer-Verlag, New York.

Peitgen, H.-O., Saupe, D. (eds) (1988) *The Science of Fractal Images,* Springer-Verlag, New York, 1988.

Peterson, M., Wegner, T. (1991) *Fractal Creations,* Waite Group Press, Mill Valley.

Pollard-Gott, L. (1986) "Fractal repetition in the poetry of Wallace Stevens," *Language and Style* **19** 233–249.

Quinn, E. (1977) *Max Ernst,* New York Graphic Society.

Roy, R., Murphey, T., Maier, T., Gills, Z., Hunt, E. (1992) "Dynamical control of a chaotic laser: experimental stabilization of a globally coupled system," *Physical Review Letters* **68** 1259–1262.

Sapoval, B. (1989) "Experimental observation of local modes in fractal drums," *Physica D* **38** 296–298.

Schiff, S., Jerger, K., Duong, D., Chang, T., Spano, M., Ditto, W. (1994) "Controlling chaos in the brain," *Nature* **370** 615–629.

Schroeder, M. (1991) *Fractals, Chaos, Power Laws: Minutes from an Infinite Paradise,* W. H. Freeman, New York.

Shearer, R. (1992) "Chaos theory and fractal geometry: their potential impact on the future of art," *Leonardo* **25** 143–152.

Shearer, R. (1995) "From Flatland to Fractaland: New Geometries in Relationship to Artistic and Scientific Revolutions," *Fractals* **3** 617–625.

Shermer, M. (1993) "The chaos of history: on a chaotic model that represents the role of contingency and necessity in historical sequences," *Nonlinear Science Today* **2** 1–13.

Sierpiński, W. (1916) "Sur une courbe cantorienne qui contient une image biunivoque et continue de toute courbe donnée," *C. R. Acad. Sci. Paris* **162** 629–632.

Stewart, I. (1989) *Does God Play Dice? The Mathematics of Chaos,* Basil Blackwell, Oxford.

Stewart, I. (1989) "Order within the chaos game?" *Dynamics Newsletter,* 4–9.

Sweet, D., Ott, E., Yorke, J. (1999) "Complex topology in chaotic scattering: a laboratory observation," *Nature* **399** 315.

Taylor, R., Michlich, A., Jones, D. (1999) "Fractal analysis of Pollock's drop paintings," *Nature* **399** 422.

Touhey, P. (1997) "Yet Another definition of chaos," *Amer. Math. Monthly* **104** 411–414.

Ulam, S. (1974) *Sets, Numbers, and Universes. Selected Papers*, MIT Press.

Ulam, S. (1976) *Adventures of a Mathematician*, Charles Scribner's Sons, New York.

Vellekoop, M., Berglund, R. (1994) "On Intervals, Transitivity = Chaos," *Amer. Math. Monthly* **101** 353–355.

Voss, R., Clarke, J. (1975) "1/f noise in music and speech," *Nature* **258** 317–318.

Voss, R., Wyatt, J. (1993) "Multifractals and the local connected fractal dimension: classification of early Chinese landscape painting," pp. 171–184 in A. Crilly, R. Earnshaw, H. Jones, eds., *Applications of Fractals and Chaos: the Shape of Things*, Springer Verlag, New York.

Waldrop, M. (1992) *Complexity: the Emerging Science at the Edge of Order and Chaos*, Simon & Schuster, New York.

Wheatley, M. (1994) *Leadership and the New Science: Learning about Organization from an Orderly Universe*, Berrett-Koehler Publishers, San Francisco.

Software

Consult the webpage

`http://www.maa.org/Fractals/welcome.html`

for an updated list.

Choate, J. (s1990) *Beginner's Fractal Exploration Toolkit*, Public Domain.

Cricket Graph III (s1994), Computer Associates International, Inc., Islandia, NY.

Georges, J., Johnson, D., Devaney, R. (s1992) *A First Course in Chaotic Dynamical Systems Software*, Addison-Wesley, Reading.

Instructional Films/Videotapes

Consult the webpage

`http://www.maa.org/Fractals/welcome.html`

for an updated list.

Clarke, A. (v1993) *Fractals. The Colors of Infinity*, Newbridge Comunications.

Devaney, R. (v1989) *Chaos, Fractals and Dynamics: Computer Experiment in Mathematics*, The Science Television Company, New York.

Devaney, R. (v1990) *Transition to Chaos*: *The Orbit Diagram and the Mandelbrot Set*, Science Television Productions, New York.

Gordon, N. (v2001) *Clouds are not Spheres*: *The Life and Work of a Maverick Mathematician*, SCETV

Hubbard, J. (v) *The Beauty and Complexity of the Mandelbrot Set*, Amer. Math. Soc., Providence.

Peitgen, H.-O., Richter, P., and Saupe, D. (v1986) *Frontiers of Chaos*, Media Magic Productions, Nicasio, Calif.

Peitgen, H.-O., Jürgens, H., Saupe, D., Zahlten, C. (v1990) *Fractals: An Animated Discussion*, W. H. Freeman, New York.

Taylor, J. (v1989) *The Strange New Science of Chaos*, Coronet Film and Video, Northbrook, IL.

URLs

`http://www.maa.org/Fractals/Welcome.html`
A basic fractal geometry course.

`http://www.maa.org/Fractals/Panorama/Welcome.html`
A growing survey of uses of fractals

`http://www.maa.org/Fractals/OCAM/welcome.html`
Web resources for Peak & Frame (1994).

`http://classes.yale.edu/math190a/Fractals/Welcome.html`
Web resources for the Yale course on fractal geometry for non-science students

`http://www.hogwildtoys.com`
Source for the "Romp" pendulum

`http://www.mathcs.sjsu.edu/faculty/rucker/chaos.htm`
Source for the "Chaos" Windows software

`http://www.chaos.umd.edu/Spheres_Photos/spheres_photo.html`
Source for fractal light scattering photographs

`http://math.bu.edu/DYSYS/dysys.html`
Boston University dynamical systems site

`http://math.bu.edu/DYSYS/arcadia`
Boston University site on fractals and chaos in *Arcadia*

`http://seawolf.uofs.edu/~monks/fractalthemes.html`
Fractal Themes site, University of Scranton

`http://www.math.utsa.edu/mirrors/maple/frame03.htm`
Maple and fractals

`http://math.albany.edu/isama/`
International Society of the Arts, Mathematics, and Architecture

`http://home.planetinternet.be/~praedsch`
A site with some nice fractal tessellations

`http://www.tessellations.com/encyclopedia.html`
Another site with some fractal tessellations

`http://www.pandromeda.com`
A site for build-them-yourself fractal planets

`http://www.math.fau.edu/Teacher/`
   `Teacher_homepage.htm`
Florida Atlantic University fractal geometry teacher enhancement site

`http://math.rice.edu/~lanius/frac/`
Rice University fractals for elementary and middle school students site

About the Editors and Other Contributors

Benoit B. Mandelbrot, best known as the author of *The Fractal Geometry of Nature*, 1982, is Sterling Professor of Mathematical Sciences at Yale University and IBM Fellow Emeritus (Physics) at the T.J. Watson Research Center. His life's work seeks a measure of order in physical, mathematical, or social phenomena that are characterized by extreme variability or roughness. The Wolf Prize cited Mandelbrot for having "changed our view of nature." Recently, his interests have broadened to include science education.

A fellow of the American Academy of Arts and Sciences and a member of the U.S. National Academy of Scences, his awards include the 1993 *Wolf Prize* for Physics, the *Barnard*, *Franklin*, *Steinmetz* and *Richardson* Medals, the *Science for Art*, *Harvey*, *Humboldt*, *Nevada*, *Honda*, and *Proctor* Prizes, the *Caltech Distinguished Service* and Scott Awards, and a number of honorary doctorates.

Trained as a topologist at Tulane University, Michael Frame became interested in fractal geometry in response to a question from a student in 1983. He has taught at several schools, most recently Union College, before moving to Yale. With physicist David Peak, he is the author of *Chaos Under Control: the Art and Science of Complexity*, a text for humanities students. Recently he has begun investigating using the web in teaching.

Melkana Brakalova
The Hotchkiss School
Lakeville, CT 06039
dral@discovernet.net

Melkana Brakalova-Trevithick holds a Ph.D. degree from Sofia University and Ed.M. degree from Harvard University. She has held positions at the Bulgarian Academy of Sciences, University of Regensburg, University of Minnesota, Washington University, American University in Bulgaria. Her research interests are in the field of quasiconformal mappings and most recently in complex dynamics. She has been a math instructor at the Hotchkiss School since 1993, where, among other advanced courses, she has enjoyed teaching, every year, a course in Fractals and Dynamical Systems. Recently, she was awarded The Arthur White Teaching Chair at Hotchkiss.

Dane Camp
New Trier High School, 385 Winnetka Avenue
Winnetka, IL 60093-4295 USA
campd@newtrier.k12.il.us

Dane Camp holds a B.A. in History (Elmhurst College), an M.S. in Mathematics (Northern Illinois University), and a Ph.D. in Historical Foundations of Education (Loyola University, Chicago), where his dissertation was "A Cultural History of Fractal Geometry: the Biography of an Idea." He has taught secondary school mathematics for over twenty years. His professional interests include the history of fractal geometry, creative methods for presenting traditional topics, and discovering connections between mathematics and other disciplines. In his free time he can often be found swimming, writing poetry, camping, singing, or scuba diving.

David Coughlin
The Hotchkiss School
Lakeville, CT 06039

David Coughlin holds a B.A. in Chemistry from Williams College and an M.S.Ed. from the University of Pennsylvania. He has been a teacher at the Hotchkiss School since 1961 where he first taught Chemistry and then moved to a full time teaching position in the Math Dept. He holds the Edward R. Tinker Chair at Hotchkiss. As someone who is virtually self-taught in mathematics, he finds his teaching intellectually challenging, but also stimulating and rewarding. His interests also include athletics, fitness, and gardening.

Donald M. Davis
Department of Mathematics, Lehigh University
Bethlehem, PA 18015 USA
dmd1@lehigh.edu

Donald Davis received his Ph.D. from Stanford University. He is Professor of Mathematics at Lehigh University, where he has taught since 1974. He has also taught at Northwestern, Johns Hopkins, and UCSD. He does research in algebraic topology and is author of *The Nature and Power of Mathematics*. He is an accomplished ultramarathon runner, having won four races of 50 miles or longer.

Robert L. Devaney
Department of Mathematics, Boston University
Boston, MA 02215 USA
bob@math.bu.edu

Robert Devaney received his Ph.D. from the University of California, Berkeley. He taught at Northwestern, Tufts, and the University of Maryland before coming to Boston University in 1980. He currently directs the Dynamical Systems and Technology Project at BU, and his main area of research is complex dynamical systems.

Vicki G. Fegers
The School Board of Broward County, Florida
600 S. E. 3rd Avenue, Fort Lauderdale, Florida, 33301
fegers_v@popmail.firn.edu

Vicki Fegers facilitates curriculum development grants for the School Board of Broward County. Her interests include providing relevant and challenging curriculum for students and authentic assessment. She received her M.S.T. in Mathematics from Florida Atlantic University.

Sandy Fillebrown
Department of Mathematics, St. Joseph's University
Philadelphia, PA 19131 USA
sfillebr@sju.edu

Sandra Fillebrown received her B.S. in Mathematics from MIT, Masters in Education from Tufts University and Ph.D. in Mathematics from Lehigh University. She taught high school mathematics for several years and has been teaching mathematics and computer science at St. Joseph's University since 1986. Her main interest outside of mathematics and education is the sport of orienteering.

Mary Beth Johnson
The School Board of Broward County, Florida
600 S. E. 3rd Avenue, Fort Lauderdale, Florida, 33301
johnsonmb@hotmail.com

Mary Beth Johnson facilitates curriculum development grants for The School Board of Broward County. Her interests include supporting professional development of teachers with an emphasis on quality instruction in the classroom. She received her M.S. in Mathematics Education from Louisiana Tech University.

Michel L. Lapidus
Department of Mathematics, University of California, Riverside
Riverside, CA 92521-0135 USA
lapidus@newmath.ucr.edu

Michel Lapidus holds Ph.D., Doctorat es Sciences, and Habilitation from the University of Paris VI. He has taught at the University of Southern California in Los Angeles, the University of Iowa, the University of Georgia in Athens, Yale University, and the University of California at Riverside, where he has been a Professor of Mathematics since 1990. He has held a number of visiting positions at universities in the US and abroad. His research interests include mathematical physics, geometric analysis, partial differential equations, dynamical systems, number theory, as well as fractal and spectral geometry. He loves reading, poetry, music, swimming, tennis, and creative teaching.

Ron Lewis
Rainbow District School Board
Ontario, Canada
ronlewis@sympatico.ca

Ron Lewis holds a B.S. in Mathematics and a Diploma in Engineering from St. Francis Xavier University, and B.Ed. from Queen's University. He teaches high school math and physics, and a self-developed course in fractals and chaos. He has presented on this topic in many workshops across North America. His special interests include gardening, camping, music, especially outdoor festivals, reading and exercising.

Kenneth Monks
Department of Mathematics, University of Scranton
Scranton, PA 18510-4699 USA
monks@uofs.edu

Ken Monks obtained his Ph.D. in algebraic topology from Lehigh University under Don Davis. He currently holds the position of Professor of Mathematics at the University of Scranton and has held visiting positions at both Lehigh and Yale Universities. When not teaching about fractals he can be found running marathons, teaching the martial arts, playing blues guitar, raising his three children, and maintaining his web site at http://www.scranton.edu/~monks (where you find out even more about him).

Brianna Murratti
Schalmont School District
Schenectady, NY
bmurratti@hotmail.com

Brianna Murratti holds a B.A. in Visual Arts from Union College, an English Special Education teacher certification and a Masters in Special Education with a concentration on Learning Disabilities, both from Southern Connecticut State University. She teaches special education and is the director of assistive technology at Schalmont School District.

David Peak
Department of Physics, Utah State University
Logan, UT 84322-4415
peakd@cc.usu.edu

David Peak holds a Ph.D. from SUNY Albany. Before moving to Utah, he was the Bailey Professor of Physics at Union College. He held a visiting position at the Institute for Advanced Study, Princeton, and was the E. Clairborne Robins Distinguished Professor of Science at the University of Richmond. His current research interests include using fractals and chaos to study biocomplexity, and the physics of dust. A strong advocate of undergraduate research, he and his students designed and built a dust experiment that flew on the space shuttle.

Douglas Ravenel
Department of Mathematics, University of Rochester
Rochester, NY 14627 USA
drav@math.rochester.edu

Douglas Ravenel has degrees from Oberlin and Brandeis. He has taught at MIT, Columbia, the University of Washington, and the University of Rochester, where he has been department chair since 1996. He has written two books and dozens of papers on algebraic topology. In 1992 he designed the college level course on fractals described in this book on the theory that it would show the aesthetic side of mathematics in a way that would attract students who would not ordinarily consider taking a nonrequired math course, and experience has shown that he was right.

Index

A Primer of Fractals and Captions:

Every book is theater and this book's plot is involved, perhaps too much so. For this reason, before the curtain is raised, the main actors are introduced on the cover. Its design, assisted by Noah Eisenkraft, combines several "icons" of fractal geometry. Some are older than the human race, others range from old to new. Together and combined with the captions, those pictures provide a primer of fractals.

Cauliflower. This real domesticated vegetable resulted from a collaboration between nature and ancient Roman gardeners. The broccoli and the cauliflower—and in particular the *romanesco* variety illustrated here—are Heaven's gift to fractalists challenged by self-styled dummies to describe what a fractal is. The whole is readily broken into florets. Each floret is a cauliflower in miniature hence is easily broken into smaller florets. Each of them, in turn, is a cauliflower in more pronounced miniature. Therefore, each part is geometrically similar to the whole, and the whole deserves to be called *self-similar*. Self-similar shapes are the simplest of all fractals.

Eruption of Mount St. Helens volcano. These billows upon billows upon billows are a striking illustration of self-similarity in a different area of nature. Many clouds in the sky provide additional examples. Using very simple fractal techniques, such shapes can be simulated—that is, imitated—in surprisingly realistic fashion. Imitation is often a necessary step toward understanding. *Credit*: M.P. Doukas & USGS, 1980.

Brownian cluster or drunkard's island. Having lost his key, a drunkard staggers along a Brownian motion as he wanders back to his starting point. The points he traverses and surrounds form a plane domain whose boundary or hull is a fractal curve. Visual and numerical observation suggested a hull fractal dimension (the first measure of roughness) equal to 4/3; the proof required 18 years of intense investigations. *Credit*: B.B. Mandelbrot & V.A. Norton, 1982.

A fractaled capital A. First, a perennial decorative design, and later an obscure bit of mathematical esoterica linked with W. Sierpinski, this construction was made into an icon of fractal geometry. Here, the self-similarity of the cauliflower has been made strict, linear.

Pharoah's breastplate. This algorithmically designed piece of jewelry is a metaphor for fractals' history, from art to art, and for the multiplicity of different contacts fractals create between fields—as represented by differently textured stones. The step from large parts of this shape to small parts does not use linear reduction, but the next least complicated geometrical construction, namely, inversion with respect to a circle. *Credit*: B.B. Mandelbrot & K.G. Monks, 1999.